QUANTITATIVE ANALYSES OF BEHAVIOR

QUANTITATIVE ANALYSES OF BEHAVIOR

Behavioral Approaches to Pattern Recognition and Concept Formation

Volume VIII

Edited by

MICHAEL L. COMMONS
RICHARD J. HERRNSTEIN
STEPHEN M. KOSSLYN
DAVID B. MUMFORD
Harvard University

LAWRENCE ERLBAUM ASSOCIATES, PUBLISHERS
1990 Hillsdale, New Jersey Hove and London

Lawrence Erlbaum Associates, Inc., Publishers
365 Broadway
Hillsdale, New Jersey 07642

Library of Congress Cataloging in Publication Data
LC card number: 86-640678
ISBN: 0-8058-0398-X

Printed in the United States of America
10 9 8 7 6 5 4 3 2 1

Volumes in the **QUANTITATIVE ANALYSES OF BEHAVIOR** series:

- Volume I: DISCRIMINATIVE PROPERTIES OF REINFORCEMENT SCHEDULES. Edited by Michael L. Commons, *Harvard University* and John A. Nevin, *University of New Hampshire*
- Volume II: MATCHING AND MAXIMIZING ACCOUNTS. Edited by Michael L. Commons, *Harvard University,* Richard J. Herrnstein, *Harvard University,* and Howard Rachlin, *State University of New York, Stony Brook*
- Volume III: ACQUISITION. Edited by Michael L. Commons, *Harvard University,* Richard J. Herrnstein, *Harvard University,* and Allan R. Wagner, *Yale University*
- Volume IV: DISCRIMINATIVE PROCESSES. Edited by Michael L. Commons, *Harvard University,* Richard J. Herrnstein, *Harvard University,* and Allan R. Wagner, *Yale University*
- Volume V: THE EFFECT OF DELAY AND OF INTERVENING EVENTS ON REINFORCEMENT VALUE. Edited by Michael L. Commons, *Harvard University,* James E. Mazur, *Harvard University,* John A. Nevin, *University of New Hampshire,* and Howard Rachlin, *State University of New York, Stony Brook*
- Volume VI: FORAGING. Edited by Michael L. Commons, *Harvard University,* Alejandro Kacelnik, *Oxford University,* and Sara J. Shettleworth, *University of Toronto*
- Volume VII: BIOLOGICAL DETERMINANTS OF REINFORCEMENT. Edited by Michael L. Commons, *Harvard University,* Russell M. Church, *Brown University,* James R. Stellar, *Northeastern University,* and Allan R. Wagner, *Yale University*
- Volume VIII: BEHAVIORAL APPROACHES TO PATTERN RECOGNITION AND CONCEPT FORMATION. Edited by Michael L. Commons, *Harvard Medical School,* Richard J. Herrnstein, *Harvard University,* Stephen M. Kosslyn, *Harvard University,* and David B. Mumford, *Harvard University*
- Volume IX: COMPUTATIONAL AND CLINICAL APPROACHES TO PATTERN RECOGNITION AND CONCEPT FORMATION. Edited by Michael L. Commons, *Harvard Medical School,* Richard J. Herrnstein, *Harvard University,* Stephen M. Kosslyn, *Harvard University,* and David B. Mumford, *Harvard University*

All volumes available from LAWRENCE ERLBAUM ASSOCIATES, PUBLISHERS

CONTENTS

ABOUT THE EDITORS

Michael L. Commons is Lecturer and Research Associate in the Department of Psychiatry at Harvard Medical School, Massachusetts Mental Health Center, and Director of the Dare Institute. He did his undergraduate work at the University of California at Berkeley, and then at Los Angeles, where in 1965 he obtained a B.A. in mathematics and psychology. In 1967 he received his M.A. and in 1973 his Ph.D. in psychology from Columbia University. Before coming to Harvard University in 1977 and becoming a Post Doctoral Fellow and then Research Associate in Psychology at Harvard University, he was an assistant professor at Northern Michigan University. He has co-edited *Quantitative Analyses of Behavior* and the annual, *Adult Development*. His area of research interest is the quantitative analysis of the construction and the understanding of reality as it develops across the life-span, especially as they affect decision processes, life-span attachment and alliance formation, ethical, social, cross-cultural, educational, legal, and private sector, applications.

Richard J. Herrnstein is Edgar Pierce Professor of Psychology at Harvard University, where he has been teaching since 1958. He obtained his A.B. in 1952 from the College of the City of New York and his Ph.D. in 1955 from Harvard University. He is past editor of *Psychological Bulletin,* has edited and authored several books, and has contributed to numerous periodicals. Among his long-standing interests is the quantitative analysis of behavior, especially with regard to the effects of reward and punishment.

Stephen M. Kosslyn has been Professor of Psychology at Harvard University since 1983. He received his Ph.D. in psychology from Stanford University in 1974. He then joined the faculty of Johns Hopkins University, where he was an Assistant Professor of developmental psychology. In 1977 he joined the faculty of Harvard University as an Associate Professor, where he taught cognitive psychology. In 1981, he received a Research Career Development Award and joined the faculty of Brandeis University. In 1982 he returned to Johns Hopkins as a Visiting Professor. Dr. Kosslyn's work focuses on visual mental imagery and high-level visual perception.

David B. Mumford has been Higgins Professor of Mathematics since 1977 and a member of the Division of Applied Science at Harvard University, where he has held a number of positions since 1961. He received his B.A. in 1957 and his Ph.D. in 1961 from Harvard University. He has also been a member of the Institute for Advanced Study, Princeton (1962) and a Visiting Professor at The University of Tokyo (1962), Tata Institute of Fundamental Research (1967, 1978) and Institut des Hautes Etudes Scientifiques, Paris (1978) as well as Nuffield Professor, University of Warwick (1970). He was elected to the National Academy of Science (1975), and is currently a MacArthur Foundation Fellow. In 1974, he received the Field Prize in mathematics. He has written numerous books and articles on mathematics, computer science and psychology.

LIST OF CONTRIBUTORS

Donald S. Blough
Department of Psychology
Brown University

John Cerella
Research and Development
Veterans Administration
Outpatient Clinic, Boston

Robert G. Cook
Sensory Science Center
Medical Center, Houston
Now at:
Department of Psychology
Tufts University

Juan D. Delius
Experimentelle Tierpsychologie
Psychologisches Institut
Ruhr-Universität, Bochum
Federal Republic of Germany
Now at:
Fachgruppe Psychologie
Allgemeine Psychologie
Universität Konstanz
Konstanz
Federal Republic of Germany

Jacky Emmerton
Experimentelle Tierpsychologie
Psychologisches Institut
Ruhr-Universität, Bochum
Federal Republic of Germany
Now at:
Sensory Sciences Center
Graduate School of Biomedical Sciences
The University of Texas Health Science Center at Houston

Pamela Jackson-Smith
Department of Psychology
University of Kentucky

Joyce A. Jagielo
Department of Psychology
University of Kentucky

Donald F. Kendrick
Department of Psychology
Middle Tennessee State University

Stephen E. G. Lea
Department of Psychology
Washington Singer Laboratory
University of Exeter

Celia M. Lombardi
Experimentelle Tierpsychologie
Psychologisches Institut
Ruhr-Universität, Bochum
Federal Republic of Germany
Now at:
Department of Biology
University of Buenos Aires

Richard Pisacreta
Department of Social Science
Ferris State University

Catriona M. E. Ryan
Department of Psychology
Washington Singer Laboratory
University of Exeter

Kenneth M. Steele
Department of Psychology
Mars Hill College

Anthony A. Wright
Sensory Sciences Center
Graduate School of Biomedical Sciences
The University of Texas Health Science Center at Houston

Thomas R. Zentall
Department of Psychology
University of Kentucky

PREFACE

The present volume is eighth in the *Quantitative Analyses of Behavior* series. We continue to seek topics in the analysis of individual behavior that have, or seem to be on the verge of, yielding theoretical formulations of a level of rigor and generality beyond the narrowly descriptive.

The contents of the present volume were first prepared for and presented at the *Eighth Symposium on Quantitative Analyses of Behavior Held at Harvard: Pattern Recognition and Concepts in Animals, People and Machines* on June 7 and 8, 1985. Subsequent revision and expansion of the papers resulted text relating to two broad sets of issues. Hence, that one year's symposium appears as two volumes. The current one, Volume VIII in the general series, presents the papers dealing with the results of experiments with animals. Volume IX addresses computational and clinical approaches to pattern recognition and concept formation.

The symposium was organized by the editors and supported in part by the Society for the Quantitative Analyses of Behavior, by the Dare Association, Inc. and the Department of Psychology at Harvard University. We owe thanks for the local arrangements to Rebecca M. Young, Patrice M. Miller and Dean Gallant, with assistance from Wilson R. Fong, Charles Ford, Peter W. Hallinan, Marjorie Bahlke Harrison, Edward Lazere, and Ann O'Brien. For help in reviewing the chapters for stylistic and organizational improvements we thank the staff of the Dare Institute.

Michael L. Commons
Harvard Medical School

I

CATEGORIES AND CONCEPTS IN BIRDS

1 Categorical Color and Shape Coding by Pigeons

Thomas R. Zentall,
Pamela Jackson-Smith, and
Joyce A. Jagielo
University of Kentucky

INTRODUCTION

A code can be thought of as an internal representation of an external stimulus. According to Lawrence (1963), a representation is a response, either overt or covert, that intervenes between the external stimulus and the overt response required by the task. Lawrence proposed that the stimulus-as-coded could consist of either some dimensional property of the external stimulus or a series of attributes needed to define the stimulus (i.e., needed to help discriminate the stimulus from other relevant stimuli). A similar view of the effective stimulus was proposed by Schoenfeld and Cumming (1963). According to Schoenfeld and Cumming, stimuli that have been associated with the same response are judged or coded as similar through mediated generalization. Thus the coded stimulus is one that is defined not by its absolute attributes but by the subject's experience with that stimulus in a particular context. The physical attributes of a stimulus may not be isomorphic with its perceptual or psychological attributes, so it may not be possible, a priori, to specify the coded stimulus in any simple way. Cumming, Berryman, and Cohen (1965) and later Carter and Werner (1978) expanded the notion of coding to include stimuli that had not been experienced. Cumming, Berryman, and Cohen (1965) trained pigeons on a matching-to-sample task in which pigeons learned to respond to the red, green, or blue comparison stimulus that matched the sample on each trial. They then replaced all the blue sample and comparison stimuli with yellow stimuli and found that the pigeons treated the yellow sample (but not the yellow comparison) as if it were red (i.e., in the presence of a yellow sample, and red and yellow comparisons, pigeons selected the red comparison). One cannot argue that Cumming and Berryman's pigeons failed to discriminate between red and yellow hues because the pigeons responded differentially to the red and yellow comparisons. Cumming and Berryman argued that the pigeons' response to the novel yellow sample was the coding response that they had learned to make in the presence of the red sample. On novel yellow-sample trials the pigeons responded to the comparison stimuli as if the sample had been red because the choice of comparison was based on the common coding response rather than the sample hue itself.

What is the coding response? Under certain conditions it may be an overt response that (a) can occur during the delay between offset of sample and onset of comparison stimuli in a delayed matching-to-sample task (Blough, 1959; Zentall, Hogan, Howard, & Moore, 1978), or (b) can be trained as differential sample behaviors (Hogan, Zentall, & Pace, 1983; Urcuioli & Honig, 1980). The observation of differential overt behavior during or following presentation of the sample stimulus provides direct evidence for the presence of coding or mediating behavior. But one can also infer coding behavior from the animal's performance on transfer tests. By providing the animal with variations of the coded stimulus and observing the animal's subsequent choice behavior, one can determine the range of stimulus values subsumed by the code and thus one can indirectly determine the defining characteristics of the code. Such a procedure is needed because the code used is likely to depend not only on the stimulus coded but also on the context in which the code is used.

When humans learn to categorize objects, for example, they often can do so at various levels of generality. A large black dog might be categorized as a labrador retriever, as a dog, as a mammal, or as an animal. The specificity of the coding response could be determined by requiring the subject to make a response in the presence of the training stimulus and asking to what degree similar responses are made to (a) other dog breeds, (b) other mammalian species, or (c) other nonmammalian species. Thus a transfer test can be used to identify the domain of the coding response.

The transfer procedure can also be applied to coding behavior by animals. When pigeons are trained to peck in the presence of natural scenes involving trees and not to peck in the presence of natural scenes in which no trees are present, the pattern of errors made to transfer scenes (and to training scenes as well) can be used to define the boundary conditions of the coding response (Herrnstein, Loveland, & Cable, 1976).

The transfer procedure can also be used to examine the coding of simpler stimuli that vary on only one or two dimensions. Wright and Cumming (1971) examined the categories pigeons used to code stimuli along the dimension of visual wavelength (hue). In their experiment pigeons were trained with a matching-to-sample task involving two hues, and then the sample hue was varied systematically between the two training values. If the hue dimension is psychologically continuous for pigeons, as it is physically continuous, one would expect the equiprobability or crossover point to depend on the two original training values (i.e., the pigeons should split the psychological distance between the two training values). Instead they found that the crossover points were absolute (i.e., the crossover points were virtually independent of the training values). Crossovers occurred either at 540 nm or at 595 nm regardless of the training values. Thus the pigeons appeared to have identified absolute boundaries between different classes of stimuli.

The phenomenon reported by Wright and Cumming (1971) is analogous to human color naming or categorization. For example, over a broad range of wavelengths humans will use the label *yellow*. At the upper boundary there will be a narrow band of wavelength that produces an ambiguous response. Beyond this point there will be another broad band of wavelengths to which humans will respond *green*.

The purpose of the experiments that follow is (a) to identify pigeons' color categories using a procedure somewhat different from that used by Wright and Cumming (1971), (b) to assess the degree to which color categories are flexible (i.e., can a particular stimulus be assigned to more than one category), (c) to determine the generality of categorical coding (i.e., is categorical coding by pigeons limited to the hue dimension), and (d) to examine pigeons' ability to use hierarchical codes (e.g., can pigeons code a red hue as *red* and also as *color*).

WITHIN-DIMENSION CATEGORICAL CODING

Experiment 1

Tasks that require the discrimination of stimuli within a category discourage the use of the category as a coding response. For humans a task that calls for differential responses to different shades of red would not encourage the use of the label *red*. On the other hand, if the task does not require discrimination between shades of red, humans might use the category label *red* to mediate the associations. The purpose of the first experiment (Zentall, Edwards, Moore, & Hogan, 1981, Experiment 1) was to provide pigeons with a task that would not discourage them from categorically coding the task stimuli. This was accomplished by training the pigeons on two, independent matching-to-sample tasks. An equal number of birds was trained on two, independent oddity-from-sample tasks. On some trials there were red and green comparison stimuli with either a red sample or a green sample. Interspersed among these trials were trials with yellow and blue comparison stimuli and either a yellow sample or a blue sample.

What the pigeons learn during this type of training can be determined by exposing the birds to test trials in which either one or the other of the comparison stimuli is replaced by either of the stimuli from the other task. The training and test trial types used in this experiment are presented in Table 1.1.

On test trials the correct comparison was either the one that was correct during

TABLE 1.1
Training and Test Trials from Experiment 1
(Zentall et al., 1981, Experiment 1)

Training	R(RG)	G(GR)	Y(YB)	B(BY)
Test				
Rule inappropriate	R(RY)	G(GB)	Y(YR)	B(BG)
	R(BG)	G(YR)	Y(GB)	B(RY)
Rule Appropriate	R(RB)	G(GY)	Y(YG)	B(BR)
	R(YG)	G(BR)	Y(RB)	B(GY)

Note. R = red, G = green, Y = yellow, B = blue

The first letter represents the color of the sample. The two letters in parentheses represent the colors of the comparisons. For the group trained with two matching tasks the first letter in parentheses represents the color defined as correct. For the group trained with two oddity tasks the second letter in parentheses represents the color defined as correct.

training or, on trials where the correct comparison was no longer present, the correct comparison was defined as the stimulus that replaced the one that was correct during training. As each of the comparison stimuli could be replaced by either of the hues from the other task, the replacement hue was counterbalanced as follows: For half the birds red replaced yellow (i.e., whenever the hue to be replaced was red, it was replaced by yellow), yellow replaced red, blue replaced green, and green replaced blue. For these birds the results showed that performance remained at a high level on all trial types (89.4% correct overall). For the remaining birds red replaced blue, blue replaced red, yellow replaced green, and green replaced yellow. For these birds test trial performance dropped to near chance levels (56.2% correct), though performance on control (training) trials remained high (90.7% correct). The drop in performance on test trials occurred regardless of whether it was the correct comparison (58.2% correct) or the incorrect comparison (54.2% correct) that was replaced. Furthermore, the effect occurred both when the two original training tasks were matching-to-sample (54.8% correct) and when the two tasks were oddity-from-sample (57.6% correct). The pigeons acted as if they had learned a common code for red and yellow samples and for blue and green samples. These codes can be described in the form of two rules: "If the sample is red/yellow respond either to the red or to the yellow comparison" and "If the sample is blue/green respond either to the blue or to the green comparison." The results are summarized in Fig. 1.1.

Rule-appropriate trial types are those to which the coding rule can be unambiguously applied. Rule-inappropriate trial types are those to which the coding rule cannot be unambiguously applied because the rule indicates either that both comparisons are correct or that neither comparison is correct.

These findings support and extend the findings reported by Cumming, Berryman, and Cohen (1965). In their experiment pigeons responded to a novel

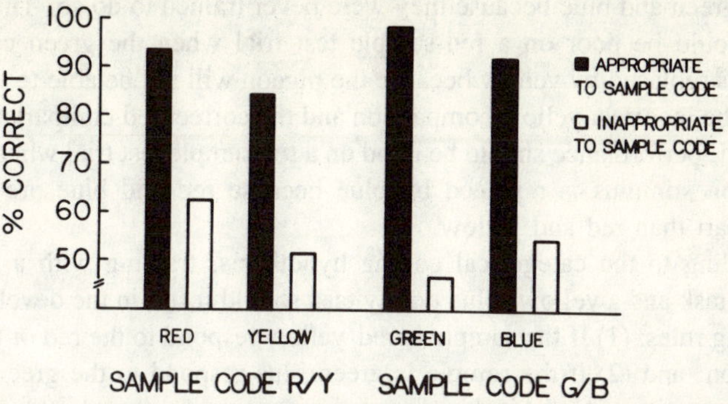

FIG. 1.1. Experiment 1: Test trial performance as a function of the appropriateness of the comparison stimuli to the sample-stimulus coding rule following training with either a red–green matching-to-sample task and a yellow–blue matching-to-sample task or a red–green oddity-from-sample task and a yellow–blue oddity-from-sample task.

yellow sample as if it were red. In our experiment they responded to familiar red and yellow samples as if they were coded the same, and also to familiar green and blue samples as if they were coded the same.

Experiment 2

Although the results of the first experiment suggest the presence of categorical color coding by pigeons, it is possible that the pigeons performed as they did because they failed to discriminate between the red and yellow hues and the blue and green hues. It may seem unlikely that stimulus values as discriminable as red and yellow and green and blue would not be discriminated, but given the fact that in original training the birds were not required to discriminate between either red and yellow, or green and blue, they may have been unable to discriminate between the members of those color pairs. Prokasy and Hall (1963) have argued that organisms will discriminate between stimuli only if they have had explicit training to do so.

The purpose of Experiment 2 (unpublished manuscript) was to rule out failure to discriminate between the red and yellow comparison stimuli and between the blue and green comparison stimuli as an explanation for the results of Experiment 1 by making one of the two training tasks a matching-to-sample task and the other training task an oddity-from-sample task. To see how this design makes it possible to distinguish between failure to discriminate and categorical coding, let us examine the predictions made following training with a red–green matching task and a yellow–blue oddity task. According to the failure-to-discriminate hypothesis, the pigeons should fail to discriminate between red and yellow and between green and blue because they were never trained to do so. Thus, performance should be poor on a red-sample test trial when the green comparison stimulus is replaced by yellow because the pigeon will not be able to distinguish between the incorrect yellow comparison and the correct red comparison. On the other hand, performance should be good on a red sample test trial when the green comparison stimulus is replaced by blue because red and blue are spectrally farther apart than red and yellow.

According to the categorical coding hypothesis, training with a red–green matching task and a yellow–blue oddity task should result in the development of two coding rules: (1) If the sample is red/yellow respond to the red or to the blue comparison, and (2) if the sample is green/blue respond to the green or to the yellow comparison. Now, if the pigeon were exposed to a red-sample test trial with a *yellow* comparison in place of the green, the pigeon would be presented the choice between red and yellow comparisons to which the red/yellow categorical coding rule (i.e., if the sample is red/yellow respond to red or blue) could be unambiguously applied and performance should remain at a high level. However, if the pigeon were exposed to a red-sample test trial with a *blue* comparison in place of the green, the red/yellow categorical coding rule could not be unambiguously applied and performance should be poor (the coding rule indicates that both comparisons are correct when in fact only the red comparison is correct). Thus for the pair of trial types described here, the discrimination failure hypothesis and the categorical coding hypothesis make opposite predictions. The same differential prediction is true for each pair of trial types produced by replacing one of the comparison stimuli by each of the stimuli from the other task. The differential prediction occurs whenever one of the tasks is matching and the other oddity.

Experiment 2 involved two training groups. Group MO was exposed to a red–green matching task and a yellow–blue oddity task. Group OM was exposed to a red–green oddity task and a yellow–blue matching task. The training and test trial types used in Experiment 2 were the same as those used in Experiment 1 (see Table 1.1). However, the reinforced comparison was not always the same in Experiment 2 as in Experiment 1 because of the mixed task design used in Experiment 2. Also, in Experiment 2 all pigeons were exposed to all the test trial types rather than to only half the trial types, as in Experiment 1. As in Experiment 1, on test trials the correct comparison stimulus was defined either as the one that was correct during training, or on trials where it was no longer present, as the one that replaced the one that was correct during training. Again the results indicated the presence of categorical coding. On test trials for which the red/yellow categorical coding rule was appropriate (i.e., could be unambiguously applied), performance (84.4% correct) was significantly higher than performance on test trials for which the red/yellow categorical coding rule was inappropriate (66.1% correct). Similarly, performance was significantly better on green/blue rule-appropriate trials (85.9% correct) than on blue/green rule-inappropriate trials (72.4% correct). The results of Experiment 2 are presented in Fig. 1.2.

The magnitude of the effect found in Experiment 2 was somewhat smaller than that found in Experiment 1. The difference in results could be due to the presence of some failure to discriminate between red and yellow and between green and blue in both experiments. In Experiment 1 generalization would have added to the effects of categorical coding, whereas in Experiment 2 generalization would have subtracted from the effects of categorical coding. If one assumes a simple additive relation between the effect of categorical coding on performance (C) and the effect of generalization on performance (G), then the difference between performance on rule-appropriate and rule-inappropriate test trials in Experiment 1 (33.2% correct) should equal the sum of the two effects, C + G, and the difference between performance on rule-appropriate and rule-inappropriate test trials in Experiment 2 (16.0% correct) should equal the difference between the two effects, C − G. Solving the two equations for C and G provides an estimate of the contribution of stimulus generalization (8.6% correct) and categorical rule learning (24.6% correct) to the present experiments. Apparently, the effect of categorical rule learning in the present experiments is almost three times as great as the effect of stimulus generalization.

Whatever the cause of the larger categorical coding effect in Experiment 1, the results of Experiment 2 support the finding that pigeons categorically code colors under conditions that preclude the pigeons' failure to discriminate among the colors.

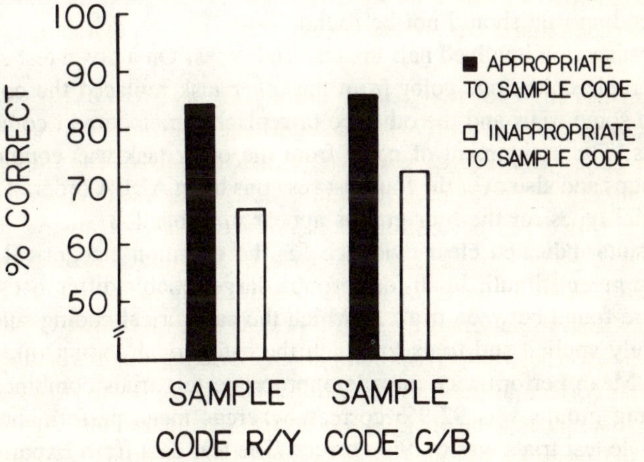

FIG. 1.2. Experiment 2: Test trial performance as a function of the appropriateness of the comparison stimuli to the sample stimulus coding rule following training with either a red–green matching task and a yellow–blue oddity task or a red–green oddity task and a yellow–blue matching task.

Experiment 3

The purpose of Experiment 3 (Zentall & Edwards, 1984) was to assess the generality of the categorical coding effects found in Experiments 1 and 2 and to examine the flexibility of the categories. In Experiment 3 each bird was trained on two tasks and the same colors were used but the colors were paired differently from the red–green and yellow–blue pairing used in Experiments 1 and 2. One of the two tasks involved red and yellow hues, the other involved green and blue hues. There were four training groups. For Group MM both tasks were matching-to-sample. For Group OO both tasks were oddity-from-sample. For Group MO the red–yellow task was matching-to-sample and the green–blue task was oddity-from-sample. For Group OM the red–yellow task was oddity-from-sample and the green–blue task was matching-to-sample.

Given the relative spectral proximity of yellow and green, it was expected that the pigeons would use a common categorical code for these two colors. Prediction of categorical coding of the red and blue stimuli was less obvious. If one assumes that the pigeon's psychological color space is circular (see Schneider, 1972), as it is in humans (Boynton, 1972, pp. 346–347), one might expect pigeons to use a common categorical code for red and blue. If the preceding assumptions are correct, the two categorical coding rules for each of the four training groups can be identified (e.g., for group MO, "If the sample is yellow/green respond to yellow or to blue" and "If the sample is red/blue respond to red or to green"). If, however, the pigeon's psychological color space is not circular or if there is a category boundary between red and blue, then evidence for a red/blue coding rule should not be found.

Each test session involved half the test trial types. On a given test session, for a given sample color, one color from the *other* task replaced the correct comparison on some trials and the other color replaced the incorrect comparison on other trials. The assignment of color from the other task was counterbalanced within groups and also over the four test sessions in an ABBA order. The training and test trial types for the four groups appear in Table 1.2.

The results indicated clear evidence for the common categorical coding of yellow and green stimuli. In all four groups, large reliable differences in performance were found between trials to which the categorical coding rule could be appropriately applied and trials to which the categorical coding rule was inappropriate. Mean performance on rule-appropriate test trials combined over the four training groups was 92.9% correct, whereas mean performance on rule-inappropriate test trials was 61.9% correct. The test data from Experiment 3 are presented in Fig. 1.3.

No evidence was found, however, for the common coding of red and blue sample stimuli. In fact, there was some evidence that red samples were coded as green and blue samples were coded as yellow, although the coding effect for red and blue samples (less than 12% difference in correct performance on rule-

TABLE 1.2
Training and Test Trials from Experiment 3
(Zentall and Edwards, 1984)

Training	Coding Rule	MM	MO	OM	OO
		Yellow Sample			
Y(YR)	Yellow/Green Rule appropriate	Y(GR)	Y(BR)	Y(YG)	Y(YB)
		Y(YB)	Y(YG)	Y(BR)	Y(GR)
	Yellow/Green Rule inappropriate	Y(BR)	Y(GR)	Y(YB)	Y(YG)
		Y(YG)	Y(YB)	Y(GR)	Y(BR)
		Green Sample			
G(GB)	Yellow/Green Rule appropriate	G(YB)	G(GY)	G(RB)	G(GR)
		G(GR)	G(RB)	G(GY)	G(YB)
	Yellow/Green Rule inappropriate	G(RB)	G(GR)	G(YB)	G(GY)
		G(GY)	G(YB)	G(GR)	G(RB)
		Red Sample			
R(RY)	Red/Blue Rule appropriate	R(BY)	R(GR)	R(RB)	R(RG)
		R(RG)	R(RB)	R(GY)	R(BY)
	Red/Blue Rule inappropriate	R(GY)	R(BY)	R(RG)	R(RB)
		R(RB)	R(RG)	R(BY)	R(GY)
		Blue Sample			
B(BG)	Red/Blue Rule appropriate	B(RG)	B(BR)	B(YG)	B(BY)
		B(BY)	B(YG)	B(BR)	B(RG)
	Red/Blue Rule inappropriate	B(YG)	B(BY)	B(RG)	B(BR)
		B(BR)	B(RG)	B(BY)	B(YG)

appropriate and rule-inappropriate trials) was considerably smaller than the coding effect for yellow and green samples (more than 30% difference in correct performance on rule-appropriate and rule-inappropriate trials). Thus to whatever degree coding of red and blue samples occurred it was relatively weak.

The results indicate that some flexibility in the development of categorical coding rules exists. The evidence that yellow can be coded as either red or green, and that green can be coded as either blue or yellow suggests that the coding categories are not biologically predetermined. It may be, however, that the specific hues selected for presentation were codable in more than one category because of their proximity to relatively fixed category boundaries. Wright and Cumming (1971) reported that pigeons performed as if they had a category boundary at about 595 nm, a point quite close to the yellow hue (dominant wave length 581 nm) used in the present experiments, and another category boundary

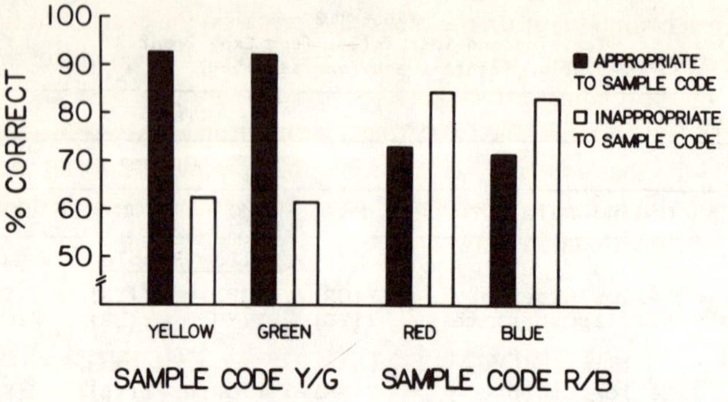

FIG. 1.3. Experiment 3: Test trial performance as a function of the appropriateness of the comparison stimuli to the sample-stimulus coding rule following training with red–yellow and blue–green matching tasks, with red–yellow and blue–green oddity tasks or with either a red–yellow matching task and a blue–green oddity task or with a red–yellow oddity task and a blue–green matching task.

at about 540 nm, a point close to the green hue (dominant wave length 526 nm). It may be that hues near the category boundaries can be assigned to either nearby category, whereas hues near the center of a category can only belong to the defining category. An example from human color categorization may be helpful in understanding this hypothesis. Many human cultures have well-defined categories for the colors blue and green. If one selects a hue at the boundary of these two colors, for example, turquoise, that hue could be placed in either category depending on the context. Following discrimination training with blue, turquoise would probably be labeled green, whereas following discrimination training with green, turquoise would probably be labeled blue.

The failure to find common categorical coding of red and blue samples can be interpreted in several different ways. First, it is possible that the psychological hue spectrum for pigeons is linear rather than circular, as is the case with humans. Schneider (1972) reported, however, that red and blue hues are psychologically closer for pigeons than would be predicted from their physical wave lengths. An alternative explanation is that the values selected for the present experiments fall far enough from each other on the pigeons' psychological color circle to preclude common coding. Finally, if Wright and Cumming (1971) are correct, the red and blue hues used in our research may simply come from distinctly different natural categories for pigeons and thus these colors would never be commonly coded. The answer to this question will have to await further research.

The results of Experiment 3 also suggest that the testing procedure may have contributed to the differences in the magnitude of coding effect found in Experiments 1 and 2. In Experiment 3 each test session included only half the test trial types (i.e., each stimulus from the other task replaced the correct comparison or the incorrect comparison but not both), and there was no difference in the magnitude of the coding effect between the two groups with the same task (Groups MM and OO) and the two groups with different tasks (Groups MO and OM). It may be that exposing the pigeons in Experiment 2 to all the test trial types during a single test session caused some confusion and washed out some of the effects of categorical coding.

DIMENSIONAL CATEGORICAL CODING

Experiment 4

The mixed-task procedure used in Experiments 2 and 3 provides a useful tool because it can be used to distinguish between categorical coding and the failure to discriminate. This tool can readily be applied to other sets of stimuli to make it possible to determine other types of categorical coding. The first three experiments dealt with coding stimuli by category within the hue dimension. In Experiment 4 (Zentall, Jackson-Smith, Jagielo, & Nallan, 1986, Experiment 1), we asked if pigeons would categorically code stimuli from the same dimension in the context of stimuli from two dimensions. For example, when presented with two line-drawn (white on black) shapes and two highly distinctive colors (red and green), would pigeons commonly code both the two shapes and the two colors?

Three questions of theoretical interest can be derived from the preceding question: (1) Are pigeons capable of categorizing stimuli other than colors? It is possible that the hue dimension represents a unique case of categorical coding because pigeons naturally group stimuli from broad regions of the visual spectrum (Wright & Cumming, 1971) based on physiological mechanisms responsible for the ability to detect color (Donner, 1953); (2) are pigeons capable of using *dimensional* categories? Will pigeons generally identify shapes as coming from a common category when the other stimuli they are presented are colors? The notion of coding by dimension is consistent with Sutherland and Mackintosh's (1971) analyzer model of discrimination learning. They suggest that dimensional analyzers are ''switched in'' as the organism is exposed to differential reinforcement for responding to stimuli along a dimension. One can think of the switching in of a dimensional analyzer as analogous to the development of a stimulus as coded; (3) are pigeons capable of hierarchical coding by category? When all the stimuli are colors pigeons appear to be able to use at least three categories to define the visual wavelength dimension. To what extent can pigeons combine color categories to define a more general ''all colors'' category when two highly dissimilar colors, that are thought to come from distinctive color categories, are presented in the context of stimuli from other dimensions (e.g., shapes)? Humans are clearly able to categorize hues by spectral region (e.g., red, yellow, green), and they are also able to name the general category to which all the hues belong (vis. color). Can pigeons use "color" as a higher order category? If they can it would suggest that they have considerable flexibility in their definition of a category.

In Experiment 4 all birds were exposed to the mixed-task design. There were four groups included in the experiment for purposes of counterbalancing: circle-red matching and plus-green oddity; circle-red oddity and plus-green matching; plus-red matching and circle-green oddity; plus-red oddity and circle-green matching. Examples of the training and test trials from Experiment 4 are presented in Table 1.3.

All the pigeons had had prior experience with a circle–plus and a red–green matching task. Test trials involved replacing either the correct or the incorrect comparison from each training trial with one or the other stimulus from the other task. If pigeons exposed to circle-red matching and plus-green oddity tasks had coded the shape samples by category, their coding rule could be described as, "If the sample is a shape respond to circle or to green." If such a coding rule had been used, good performance would be expected both on test trials with circle and plus as comparisons and on those with red and green as comparisons because the coding rule could be appropriately applied. Conversely, one would expect performance to be poor both on test trials with circle and green as comparisons and on those trials with plus and red as comparisons because the coding rule would be inappropriate to the comparison choice (i.e., the coding rule indicates that either both comparisons or neither comparison is correct).

The results indicated that performance on rule-appropriate shape-sample test trials (83.0% correct) was significantly better than performance on rule-inappropriate shape-sample test trials (72.3% correct). The test data for Experiment 4 are presented in Fig. 1.4.

The most convincing evidence for shape-sample coding by category comes from the pigeons' performance on test trials that involved only shapes. According to the discrimination-failure hypothesis, performance on trials involving only shapes should be poor both because they are relatively similar and because the pigeons were not trained to discriminate between the two shapes. According to the coding hypothesis, however, performance should be good on these trials as long as the coding rule can be unambigously applied. Examination of the results indicated that performance on test trials that involved only shapes (73.3% correct) was significantly better than performance on comparable test trials that involved one shape and one color as comparison stimuli (49.0% correct). Thus the results of Experiment 4 provide clear evidence for the coding of shape samples by category.

The color-sample test results, on the other hand, were not as predicted. In fact, performance on rule-appropriate color-sample test trials (69.5% correct) was significantly worse than performance on rule-inappropriate color-sample test trials (86.5% correct). The most reasonable interpretation of the color-sample

TABLE 1.3
Potential Coding Rules and Training and Test Trials for the Plus-Red Matching Group in Experiment 4
(Zentall et al., 1986, Experiment 1)

	Matching		Oddity	
Training	+(+R)	R(R+)	O(GO)	G(OG)
Test				
Rule inappropriate	+(+G)	R(RO)	O(G+)	G(OR)
	+(OR)	R(G+)	O(RO)	G(+G)
Rule appropriate	+(+O)	R(RG)	O(GR)	G(O+)
	+(GR)	R(O+)	O(+O)	G(RG)

Coding Rules: (1) "If the sample is a shape respond to the plus or to the green comparison."
(2) "If the sample is a color respond to the red or to the circle comparison."

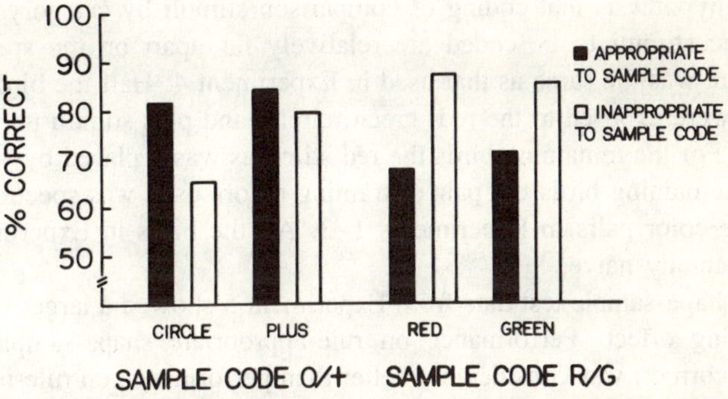

FIG. 1.4. Experiment 4: Test trial performance as a function of the appropriateness of the comparison stimuli to the sample-stimulus coding rule following training with circle-red matching and plus-green oddity, circle-red oddity and plus-green matching, plus-red matching and circle-green oddity, or plus-red oddity and circle-green matching.

data is that they suggest a comparison coding effect. For the two, red-matching groups, for example, the coding rule would be, "If the sample is red respond to the color comparison." The rule would work quite well when the correct red comparison was replaced by green, but it would provide an ambiguous instruction when the incorrect shape comparison was replaced by green (i.e., both comparisons would be colors). Earlier in this chapter we argued that results such as those just described for color-sample test trials suggest that the birds were not discriminating between the stimuli (in this case red and green). But given the fact that under similar conditions pigeons can readily discriminate between red and yellow (Zentall et al., 1981) and between green and yellow (Zentall & Edwards, 1984), failure to discriminate red from green is quite unlikely.

Why did comparison coding occur in Experiment 4, but not in earlier experiments? First, in Experiment 4 each training trial had one color and one shape as comparison stimuli. The high discriminability of these stimuli may have resulted in a discrimination based on general characteristics of the stimuli (e.g., dimensional categorization) rather the specific absolute properties of the stimuli.

Second, in Experiment 4 the colors to be coded (red and green) were farther apart spectrally than the colors in the earlier research. When the two tasks involved all four colors, the pigeons could code red as yellow or could code yellow as green, but in the present research the only color coding that could occur was red as green. It is possible that sample coding will occur only when the colors are relatively close spectrally.

Experiment 5

The main purpose of Experiment 5 (Zentall et al., 1986, Experiment 2) was to test the hypothesis that coding of comparison stimuli by category occurs only when the stimuli to be coded are relatively far apart on the spectrum. The procedure was the same as that used in Experiment 4. Half the birds in Experiment 5 were exposed to the red, green, circle, and plus stimuli used in Experiment 4. For the remaining birds the red stimulus was replaced by yellow. Thus for the remaining birds the pair of training colors used was spectrally close as were the color pairs in Experiments 1–3. All the birds in Experiment 5 were experimentally naive.

The shape-sample test data from Experiment 5 showed a large, reliable sample coding effect. Performance on rule-appropriate shape-sample test trials (93.2% correct) was considerably better than performance on rule-inappropriate shape-sample test trials (60.2% correct). The shape-sample data are presented in Fig. 1.5. The magnitude of the shape-sample coding effect in Experiment 5 (33.0% correct) was substantially greater than the shape-sample coding effect in Experiment 4 (10.7% correct). The difference can be attributed to the fact that the pigeons in Experiment 4 had learned to discriminate the circle and plus shapes prior to being exposed to the present training procedures, whereas the pigeons in Experiment 5 were experimentally naive. Apparently, prior discrimination training can reduce an organism's tendency to commonly code those stimuli.

Analysis of the color-sample test data also generated results similar to the results of Experiment 4. Overall, performance on color-sample, rule-appropriate test trials (65.8% correct) was significantly worse than performance on color-sample, rule-inappropriate test trials (81.8% correct). Thus the comparison coding effect for color samples appears to be reliable. The color-sample test data also appear in Fig. 1.5. Notice in the figure that comparison coding occurs in this context whether the training colors are red and green (top of Fig. 1.2) or yellow and green (bottom of Fig. 1.2). Apparently it is the fact that the comparison

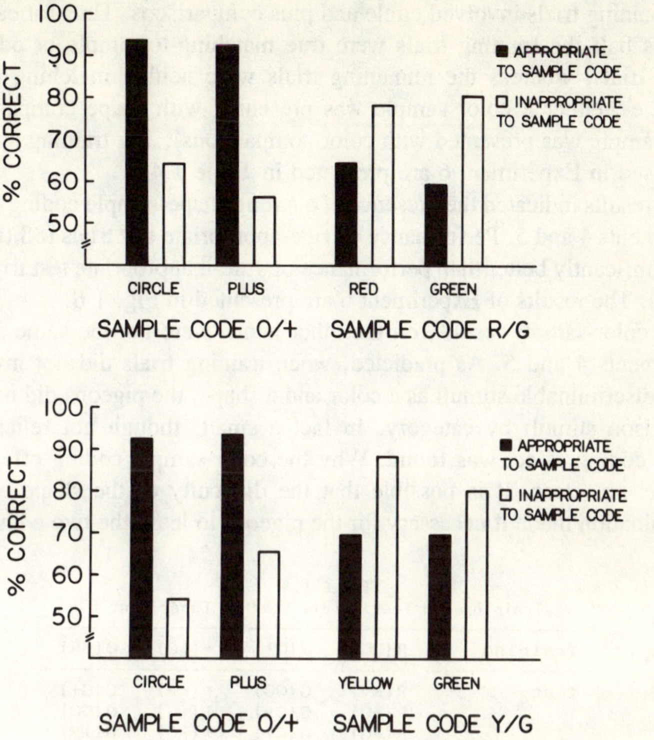

FIG. 1.5. Experiment 5: Test trial performance as a function of the appropriateness of the comparison stimuli to a sample stimulus coding rule following training with circle-red matching and plus-green oddity, circle-red oddity and plus-green matching (top), circle-yellow matching and plus-green oddity, or circle-yellow oddity and plus-green matching (bottom).

stimuli are easy to discriminate (one shape, one color) that is responsible for the coding of comparison stimuli by category and not the spectral distance between the color stimuli.

Experiment 6

If, in fact, high discriminability of comparison stimuli (e.g., one color and one shape) is responsible for coding of comparison stimuli by category, then one should be able to eliminate comparison coding by exposing pigeons to a training task that does not involve the presentation of trials with one shape and one color comparison stimuli. In experiment 6 all trials involved the presentation of comparison stimuli from the same dimension. Half the trials with each of the four sample stimuli (red, green, circle, plus) involved red and green comparisons, and the remaining trials involved circle and plus comparisons. Thus for each sample stimulus half the training trials were true matching-to-sample or oddity-from-sample trials, whereas the remaining trials were neither matching nor oddity trials (i.e., either a color sample was presented with shape comparisons or a shape sample was presented with color comparisons). The training and test trial types used in Experiment 6 are presented in Table 1.4.

The results indicated the presence of a strong shape-sample coding effect as in Experiments 4 and 5. Performance on rule-appropriate test trials (88.0% correct) was significantly better than performance on rule-inappropriate test trials (56.5% correct). The results of Experiment 6 are presented in Fig. 1.6.

The color-sample results, on the other hand, were not the same as those of Experiments 4 and 5. As predicted, when training trials did not involve such highly discriminable stimuli as a color and a shape, the pigeons did not code the comparison stimuli by category. In fact a small, though not reliable, color-sample coding effect was found. Why the color-sample coding effect was not larger is not clear. It is possible that the difficulty of the shape comparison discrimination made it necessary for the pigeons to learn the two components of

TABLE 1.4
Training and Test Trials Used in Experiment 6

Training	R(RG)	G(O+)	+(+O)	O(GR)
Test	R(R+)	G(OG)	+(+R)	O(G+)
	R(RO)	G(OR)	+(+G)	O(GO)
	R(+G)	G(G+)	+(RO)	O(+R)
	R(OG)	G(R+)	+(GO)	O(OR)

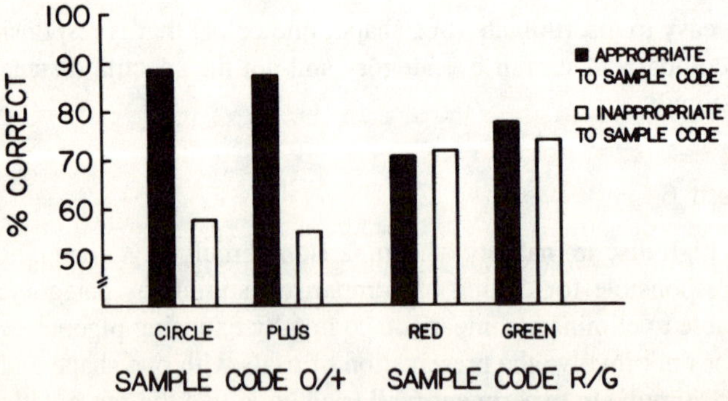

FIG. 1.6. Experiment 6: Test trial performance as a function of the appropriateness of the comparison stimuli to a sample coding rule following conditional discrimination training with red, green, circle, and plus stimuli (see text).

the color-sample training task at different rates and thus reduced their tendency to code the color samples by category. Such an explanation does not account for the strong shape-sample categorical coding effect, however. In any event, it is clear that the coding of color comparisons by category can be disrupted by using less discriminable comparison stimuli.

CONCLUSIONS

The research described in this chapter has a number of implications for research in animal cognition. First, taken as a whole, the methodology presented in Experiments 2–5 provides a useful tool for the study of coding by category in the absence of discrimination-failure effects. This methodology can be applied to any stimulus set in order to investigate both intradimensional and dimensional categorization. In addition, this research has theoretical implications.

Intradimensional Categories

The first three experiments provide evidence that there are (at least) three psychologically meaningful color categories for the pigeon. One category includes all the colors our culture calls red and orange and some of the colors we call yellow. A second category includes the rest of the yellows and some of the greens. A third category includes the rest of the greens and the blues. Although these categories do not correspond to the major categories used by humans, they do correspond to the three color categories found for pigeons by Wright and Cumming (1971). The fact that pigeons sometimes categorize yellow as red and sometimes categorize yellow as green suggests that the category boundaries are somewhat flexible. It may be, however, that the flexibility is limited to hues that are relatively close in wavelength to the pigeon's "natural" color boundaries.

The absence of common coding for red and blue colors in Experiment 3 has potential implications for the shape of the underlying psychological (psychophysical) hue dimension for pigeons. It is possible that the pigeon views the underlying dimension as linear rather than circular as in the case of humans (but see Schneider, 1972). Alternatively, it might be that the particular values of red and blue were not sufficiently far apart spectrally to put them close enough on the psychological color circle to produce common coding. Finally, it is possible that, although the psychological color dimension for the pigeon is circular, there may be a natural categorical boundary between red and blue hues.

Dimensional Categories

The results of Experiments 4, 5, and 6 indicate that pigeons can code line-drawn shapes as belonging to a common category. It could be argued that shape is not the only label that one could give to the common code used to categorize these black and white stimuli, but it is clear that the coding process observed does not represent a failure to discriminate between these stimuli.

The results of the last three experiments also suggest the presence of a general category-based code for colors. This general color code, together with the specific intradimensional color categories, suggest the flexibility of a hierarchical coding system that depends on the characteristics of the other task-relevant stimuli. An unexpected finding was that the comparison stimuli rather than the samples were coded as *colors*. The results of Experiment 5 indicated, however, that the coding of comparison stimuli was not an artifact of the specific colors selected. Instead, it appears that this effect was produced by the relative ease of discriminability of the comparison stimuli, as shown by the results of Experiment 6.

A number of important questions remain to be answered. With regard to the intradimensional color coding effects: (a) To what extent are the color categories flexible? Are they flexible only when wavelengths close to the category boundaries found by Wright and Cumming (1971) are selected, or are the categories even more flexible and determined by whatever context is provided by the other training hues? (b) Is the pigeon's psychological color space circular as suggested by Schneider (1972), or is it linear as suggested by the present research? If it is circular, do pigeons have a color category that consists of both long wavelength reds and short wavelength blues, or is there a natural boundary between the reds and the blues? (c) Is the dimensional coding of colors by category found in Experiments 4 and 5 limited to comparison coding, or are there conditions under which similar sample coding effects can be found?

The procedures described in this research generate a variety of other questions about coding by category. For example, are there intradimensional categories for shapes (e.g., curved vs. straight line shapes, same shape rotated vs. different shape rotated, mirror image of same shape vs. mirror image of different shape, low vs. high "center of gravity," etc.)? Are there other dimensions that pigeons can divide into categories? To what extent can pigeons use dimensional categories other than colors and shapes? The methodology described in this research can help to provide answers to these and other questions.

ACKNOWLEDGMENTS

The research reported in this chapter was supported by grants from the National Institute of Mental Health (MH-24092 and MH-35378), the National Science Foundation (BSN-8418275), and the University of Kentucky Research Foundation. We thank David E. Hogan and Charles A. Edwards for their contribution to the research reported here.

REFERENCES

Blough, D. S. (1959). Delayed matching in the pigeon. *Journal of the Experimental Analysis of Behavior, 2,* 151–160.

Boynton, R. M. (1972). Color vision. In J. W. Kling & L. A. Riggs (Eds.), *Woodworth and Schlosberg's experimental psychology* (Vol. 1, pp. 315–368). New York: Holt, Rinehart, & Winston.

Carter, D. E., & Werner, T. J. (1978). Complex learning and information processing by pigeons: A critical analysis. *Journal of the Experimental Analysis of Behavior, 29,* 565–601.

Cumming, W. W., Berryman, R., & Cohen, L. R. (1965). Acquisition and transfer of zero-delay matching. *Psychological Reports, 17,* 434–445.

Donner, K. O. (1953). The spectral sensitivity of the pigeon's retinal elements. *Journal of Physiology, 122,* 524–537.

Herrnstein, R. J., Loveland, D. H., & Cable, C. (1976). Natural concepts in pigeons. *Journal of Experimental Psychology: Animal Behavior Processes, 2,* 285–302.

Hogan, D. E., Zentall, T. R., & Pace, G. (1983). Control of pigeons' matching-to-sample performance by differential sample response requirements. *American Journal of Psychology, 96,* 37–49.

Lawrence, D. H. (1963). The nature of a stimulus: Some relationships between learning and perception. In S. Koch (Ed.), *Psychology: A study of a science* (Vol. 5, pp. 179–212). New York: McGraw-Hill.

Prokasy, W. F., & Hall, J. F. (1963). Primary stimulus generalization. *Psychological Review, 70,* 310–322.

Schneider, B. (1972). Multidimensional scaling of color difference in the pigeon. *Perception & Psychophysics, 12,* 373–378.

Schoenfeld, W. N., & Cumming, W. W. (1963). Behavior and perception. In S. Koch (Ed.), *Psychology: A Study of a Science* (Vol. 5, pp. 213–252). New York: McGraw-Hill.

Sutherland, N.S., & Mackintosh, N. J. (1971). *Mechanisms of animal discrimination learning.* New York: Academic Press.

Urcuioli, P. J., & Honig, W. K. (1980). Control of choice in conditional discriminations by sample-specific behaviors. *Journal of Experimental Psychology: Animal Behavior Processes, 6,* 251–277.

Wright, A. A., & Cumming, W. W. (1971). Color-naming functions for the pigeon. *Journal of the Experimental Analysis of Behavior, 15,* 7–17.

Zentall, T. R., & Edwards, C. A. (1984). Categorical color coding by pigeons. *Animal Learning and Behavior, 12,* 249–255.

Zentall, T. R., Edwards, C. A., Moore, B. S., & Hogan, D. E. (1981). Identity: The basis for both matching and oddity learning in pigeons. *Journal of Experimental Psychology: Animal Behavior Processes, 7,* 70–86.

Zentall, T. R., Hogan, D. E., Howard, M. M., & Moore, B. S. (1978). Delayed matching in the pigeon: Effect on performance of sample-specific observing responses and differential delay behavior. *Learning and Motivation, 9,* 202–218.

Zentall, T. R., Jackson-Smith, P., Jagielo, J. A., & Nallan, G. B. (1986). Categorical shape and color coding by pigeons. *Journal of Experimental Psychology: Animal Behavior Processes, 12,* 153–159.

2 Rudimentary Rule-Governed Behavior in the Pigeon

Richard Pisacreta
Ferris State University

The researchers in epistemology, including psychologists and developers of artificial-intelligence models, have frequently proposed that one of the primary indices of intelligence is the capacity to learn concepts. One type of concept learning in animals is inferred when the animal's behavior suggests that it has learned some rudimentary rule involving same/different judgments; that is, an animal's behavior is controlled to some extent by conditional relationships among stimuli rather than by their absolute values (Premack, 1978). Failure to produce evidence of concept learning in experimental situations does not necessarily mean that an animal is incapable of it.

Herrnstein and Loveland (1964) wrote that the pigeon will form a complex concept when placed in a situation that demands one. It has been one of the challenges for epistemological research over the last two decades—to employ procedures that reinforce the animal for attending to the relational factors among stimuli, for example, whether they are the same or different, rather than attending only to the absolute properties of the stimuli.

It is well known that, when pigeons serve as the subjects, the standard three key matching-to-sample procedure (MTS) employing three hues or forms usually fails to produce convincing evidence of concept learning when a fourth sample is introduced (Carter & Werner, 1978; Cumming & Berryman, 1961; Farthing & Opuda, 1974). It is believed that this occurs because the bird is not required to learn a matching concept. The bird can perform well above chance by just learning a series of two component chains (e.g., if red is the sample then peck the red comparison stimulus), or by learning three key discriminations, e.g., given the key configuration—red, red, blue; peck red (Carter & Werner, 1978).

Facilitated matching performance refers to experimental situations in which the bird shows a "savings in learning" (Premack, 1978); that is, the pigeon performs above chance, or at least makes fewer errors when exposed to novel MTS problems relative to its acquisition performance with the original stimuli. The demonstrations of facilitated matching that have been reported have often used modified MTS procedures designed to reduce proactive interference, or to minimize the possibility that the bird would learn only two component chains or three key discriminations. Several papers have suggested that facilitated matching performance can be improved by employing a large number of training stimuli (e.g., Holmes, 1979; Lombardi, Fachinelli, & Delius, 1984; Wright, Santiago, Urcuioli, & Sands, 1983). For example, Lombardi et al. reported that patterns transferred better when pigeons were trained to match 20 visual patterns relative to other birds that had been initially trained with only five samples. They suggested that employing several sample stimuli teaches the bird to learn a more economical general matching rule. Presenting complex stimuli instead of simple hues also appears to facilitate transfer of matching performance (e.g., Cerella, 1979; Delius & Nowak, 1982; Dittrich, this volume).

Zentall, Edwards, and Hogan (1983) suggested that providing "negative instance trials," that is, trials with two nonmatching comparison stimuli, will also facilitate transfer of matching behavior. They trained pigeons to match a circle and a plus. During a third of the trials neither comparison stimulus matched the sample (e.g., if the sample was a plus, both comparison stimuli were circles). The trial advanced after the bird had not pecked a comparison stimulus for 3 seconds. When red and green stimuli were introduced, these birds produced better facilitated matching transfer relative to other pigeons trained with the same stimuli but not exposed to negative instance trials.

Other manipulations have included sequential, rather than simultaneous, presentation of the comparison stimuli (Urcuioli & Nevin, 1975); and increasing the number of incorrect comparison stimuli available during each trial (Zentall, Hogan, & Edwards, 1980). These procedures were expected to facilitate matching performance by teaching S− as well as S+ relationships among sample and comparison stimuli.

These papers suggested that situations should be provided that would be so complex that the animal should learn a rule in order to produce reinforcers economically (cf. Herrnstein & Loveland, 1964; Pisacreta, 1982; Pisacreta & Witt, 1983). It was believed that if each of these five manipulations individually contributed to facilitated matching performance, then a procedure that incorporated all of them might produce even better transfer results.

INITIAL STUDIES

In the experiments reported here, an attempt was made to teach pigeons what Lea and Ryan (1983) refer to as a formal concept, "sameness," as opposed to a "perceptual" concept, such as the presence or absence of a particular stimulus class (see Cook, Wright, & Kendrick, this volume).

In the first experiment (Pisacreta, Redwood, & Witt, 1984), pigeons were trained on a six-key sequential matching-to-sample procedure (SMTS). The response panel used in this first and subsequent experiments had nine pecking keys arranged in a 3 by 3 matrix. Stimuli were rear projected on each of the 2.7- cm (diameter) keys by inline projectors. For reference, the keys are numbered from upper left (1) across to lower right (9). The feeder aperture was located directly below key 8 and provided 3-second access to grain. Figure 2.1 displays the stimuli that were employed as stimuli set 1. These figure stimuli were used instead of hues or forms to reduce the viability of explaining facilitated matching performance on the basis of stimulus-generalization mechanisms (e.g., wave length). Figure stimuli cannot be easily ordered along some dimension that facilitates generalization.

After the preliminary training procedures, a trial began with one of five randomly chosen samples (a woman, bird, hand, beetle, or flower) projected on key 5, the sample key. The third peck on the sample key (fixed ratio 3, FR3) produced only one comparison stimulus, on key 6. Every third additional sample peck produced another comparison stimulus, up to a maximum of five comparison stimuli. The comparison keys were always lighted in the sequence 6, 4, 1, 2, and 3. A peck on the matching comparison stimulus produced grain, an intertrial interval, and the next trial. A ''correction'' procedure eliminated the comparison stimuli whenever an error occurred and repeated the same trial until a match was produced. After acquisition (80% accuracy), the birds showed facilitated matching transfer to a fish, a rabbit, and a red hue. Two of the birds yielded

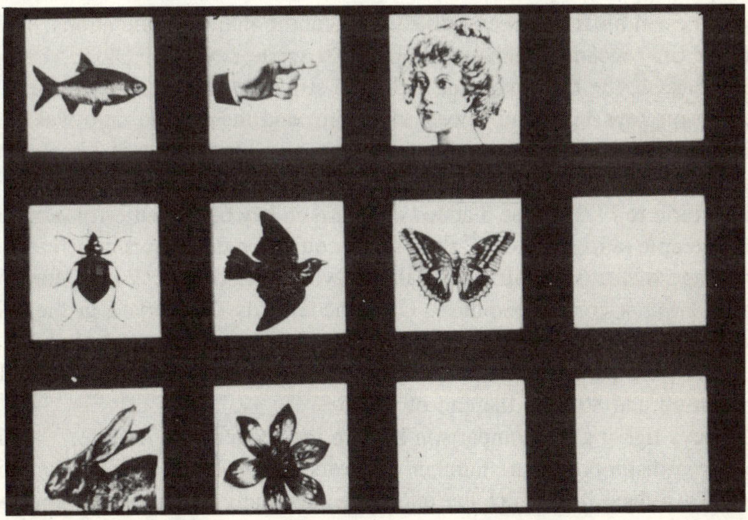

FIG. 2.1. The figure stimuli presented as samples and comparison stimuli during phases 1–6 inclusive (from *Harter's picture archive for collage and illustration,* by J. Harter (Ed.), 1978, New York, Dover).

only a 10–15% reduction in matching accuracy during the first few sessions that presented these novel stimuli. Recovery of matching accuracy levels occurred within 6 to 10 sessions.

In the second experiment, Pisacreta, Lefave, Lesneski, and Potter (1985) trained pigeons on the sequential oddity-from-sample (SOFS) version of the procedure. The birds were trained to peck the sample key (which presented 1 of 3 color drawings of different species of birds), until a nonmatching comparison stimulus was presented on 1 of the 5 other keys. After acquisition (70–90% accuracies) the 3-bird stimuli were replaced by a new set of bird drawings. The succeeding 12 phases of the study introduced novel sets of bird, butterfly, and human-face stimuli. Both birds showed transfer of oddity learning to the new samples.

In this experiment, Pisacreta and his associates expanded on this SMTS procedure by introducing five or nine novel samples during consecutive phases. An attempt has also been made to assess the importance of sequential versus simultaneous presentation of the comparison stimuli, as well as various ways of introducing novel samples.

METHOD

Subjects

Birds 1 and 2 had previously been trained on the standard 3-key MTS procedure with the stimuli depicted in Fig. 2.1. Pigeons 3 and 4 were used in the first experiment (Pisacreta, Redwood, & Witt, 1984), but it had been 15 months since any of these 4 birds had been trained with conditional discriminations. The fifth bird had only recently been autoshaped to a white key.

In phase 1, the bird, woman, and hand stimuli of Fig. 2.1 were used. Only comparison keys 1, 2, and 3 were available, and the sample ratio was 1. Each sample peck lighted another comparison key up to a maximum of 3 keys. Phase 2 introduced another comparison key (key 4), the beetle stimulus, and raised the sample ratio to FR2. Phase 3 added comparison key 6, the butterfly sample, and a FR3 sample ratio. Figure 2.2 shows the acquisition data. Accuracy refers to the percentage of the 36 reinforced trials for which the first peck on a comparison stimulus was a correct response. Only the 2 birds that served in the original experiment performed above chance during phase 1. By the end of phase 2, all the birds were performing above chance and were maintaining accuracy levels between 60 and 80% by the end of phase 3.

Always lighting the comparison keys in the sequence 6, 4, 1, 2, 3 tended to produce a disproportionate number of errors on the first 2 comparison keys to light. To reduce this effect the experimenters started lighting the comparison keys in random-order sequences during phase 4. In phase 5, a zero-delay condition was introduced to replicate some of the results of the earlier experiment. At

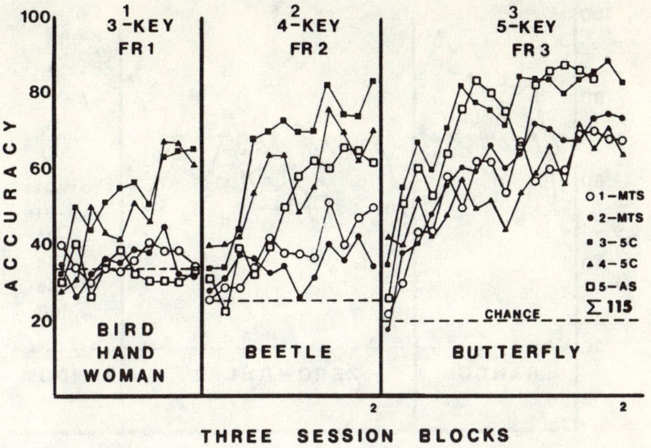

FIG. 2.2. The percentage of the 36 reinforced session trials in which the first peck on a comparison stimulus was a match. The phase, number of comparison keys, and sample ratio are noted at the top of each panel. The experimental history of the birds included training on matching-to-sample (MTS), the five-choice procedure (5C), or autoshaping (AS). Σ indicates cumulative experimental sessions.

trial onset key 5 presented the sample stimulus as before. The third sample peck produced a comparison stimulus but also replaced the sample stimulus with white light. Each additional peck on key 5 replaced white with the sample stimulus, but only for as long as the key was depressed. The sample was replaced with white light after each peck. Therefore the sample was absent when the bird pecked a comparison stimulus.

Figure 2.3 shows that the birds continued to perform well above chance, although their accuracy levels decreased about 20% during the zero-delay condition, presumably because they increasingly had to use "working memory" of the sample (Honig, 1978; Roitblat, 1984). Phase 6 restored the simultaneous MTS conditions of phase 4 and produced an immediate improvement in accuracy levels, as expected. Bird 4 became ill and was retired at the end of this phase.

Sample Replacement Versus Multiple Trials

It was then decided to explore different ways of introducing novel samples. They included totally replacing familiar sample and comparison stimuli with new ones, or having novel matching problems presented during only half the session trials. Two new sets of stimuli were employed, and they are presented in Fig. 2.4. For reference, the stimuli are numbered from upper left (1), across and down to lower right (12). During phase 7, birds 1 and 5 were trained to match the 5 novel stimuli of stimuli set 2, (1, 2, 5, 6, 10), whereas birds 2 and 3 were trained with stimuli set 3 (3, 7, 8, 9, 11). After 40 sessions of training, a

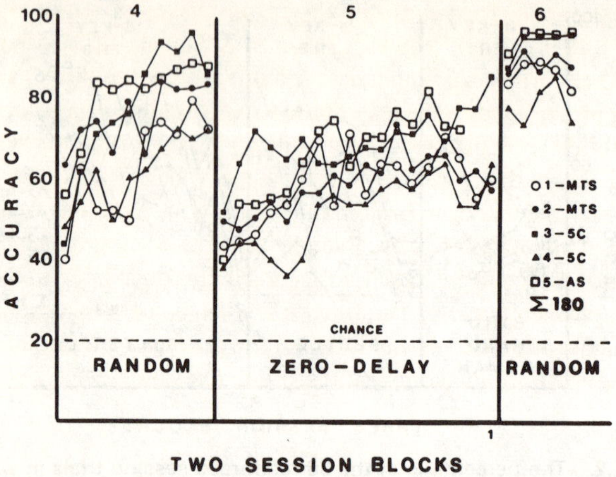

FIG. 2.3. Matching accuracies produced during phases 4–6. The comparison keys were lighted in random sequences across trials. The sample stimulus was replaced with white light after each sample peck during the zero-delay condition.

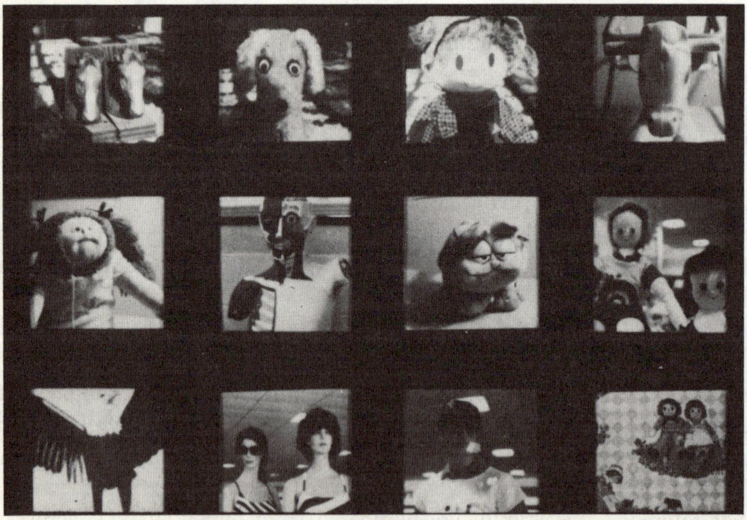

FIG. 2.4. The figure stimuli employed during phases 7 and 8.

multiple-matching condition was introduced, in phase 8 (D'Amato & Salmon, 1984; Fujita, 1983; Zentall, Edwards, Moore, & Hogan, 1981); that is, during half the reinforced trials, the familiar set of stimuli that the birds had learned to match during phase 7 was replaced with the five stimuli of the alternate stimuli set. Thus the birds were matching 10 stimuli per session, 5 familiar and 5 novel.

Figure 2.5 shows the acquisition data generated during phases 7 and 8. The filled circles indicate that each bird was matching above chance early in phase 7 and stabilized at about 75–90% accuracy levels. Introducing the novel stimuli in phase 8 had little effect on the matching accuracies produced during the trials with familiar stimuli (open circles). The filled triangles indicate that each bird generated steeper acquisition curves and required fewer sessions to produce matching levels with the novel stimuli comparable to those matching levels yielded in phase

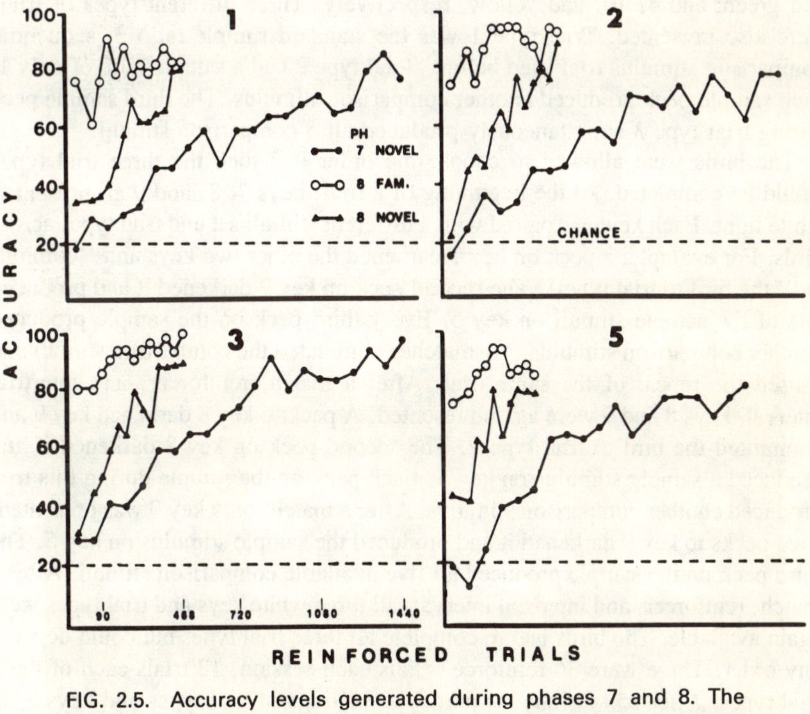

FIG. 2.5. Accuracy levels generated during phases 7 and 8. The number above each panel represents the bird. Filled circles represent acquisition data generated while pigeons learned to match the five novel stimuli during phase 7. Open circles show matching levels produced with these now (FAM)iliar stimuli during phase 8. Filled triangles show the acquisition curves generated while pigeons learned to match the five novel stimuli during phase 8.

7. This occurred even though the novel stimuli were presented during only half the reinforced trials of each session of phase 8. A subjects by treatments analysis of variance of the differences in acquisition rates between phases 7 and 8 was statistically significant by the second session, $F(1,3) = 18.69, p < .025$ (group data).

Sequential Versus Simultaneous Comparison Stimuli

At this point, the experimenters attempted to assess the importance of sequential versus simultaneous presentation of the comparison stimuli as well as the significance of the sample-ratio value. Three new sets of stimuli were employed and they are presented in Fig. 2.6. Each stimuli set had 3 stimuli, 2 human figures and a color. Stimuli set 4, 5, and 6 consisted of pictures: 3, 11, and blue; 7, 8, and green; and 4, 10, and yellow, respectively. Three different types of trials were also presented. Trial type 1 was the standard sample ratio 3, sequential comparison stimulus trial used before. Trial type 2 had a sample ratio of only 1, each sample peck produced another comparison stimulus. The third sample peck during trial type 3 simultaneously produced all 5 comparison stimuli.

The birds were allowed to choose the order in which the three trial types would be completed. At the beginning of a trial, keys 7, 8, and 9 all presented white light. Each key was paired with a different stimuli set and trial type, across birds. For example, a peck on key 7 darkened the other two keys and "committed" the bird to trial type 1. The second peck on key 7 darkened it and produced one of the sample stimuli on key 5. Every third peck on the sample produced another comparison stimulus. Mismatches eliminated the comparison stimuli and caused the repeat of the same trial. After a match, reinforcer, and intertrial interval, keys 8 and 9 were again presented. A peck to key 8 darkened key 9 and committed the bird to trial type 2. The second peck on key 8 darkened it and produced a sample stimulus on key 5. Each peck on the sample during this trial produced another comparison stimulus. After a match, only key 9 was presented. Two pecks to key 9 darkened it and produced the sample stimulus on key 5. The third peck on the sample produced all five available comparison stimuli. After a match, reinforcer, and intertrial interval, all three white keys and trial types were again available. The birds had to complete all three trial types but could do so in any order. There were 36 reinforced trials each session, 12 trials each of the 3 trial types. After 25 sessions the particular pairings of trial types and keys 7, 8, and 9 were changed to insure that it was the particular trial types that were responsible for the sequences that emerged.

Figure 2.7 shows that trial type 1 produced the best accuracy levels during the first 25 sessions of phase 9. Simultaneous presentation of the comparison stimuli, trial type 3, produced the poorest acquisition curve. A subjects-by-treatments ANOVA was statistically significant, $F(2,6) = 8.482, p < .025$ (group data).

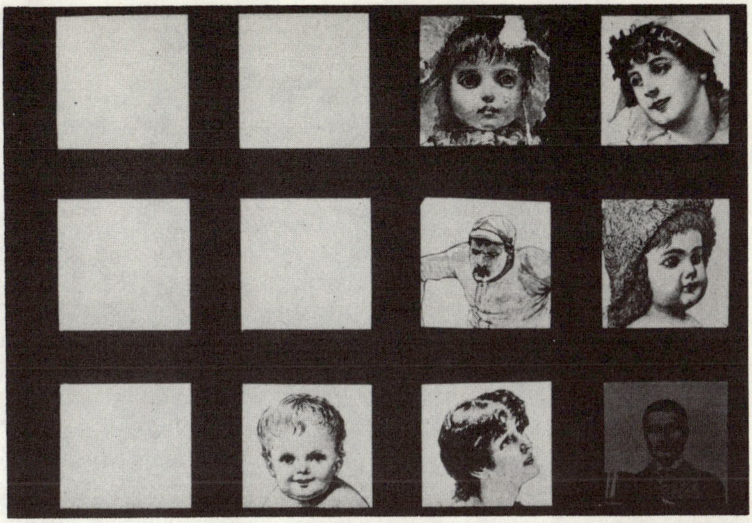

FIG. 2.6. The stimuli presented during phase 9. The blank squares were capable of presenting different hues.

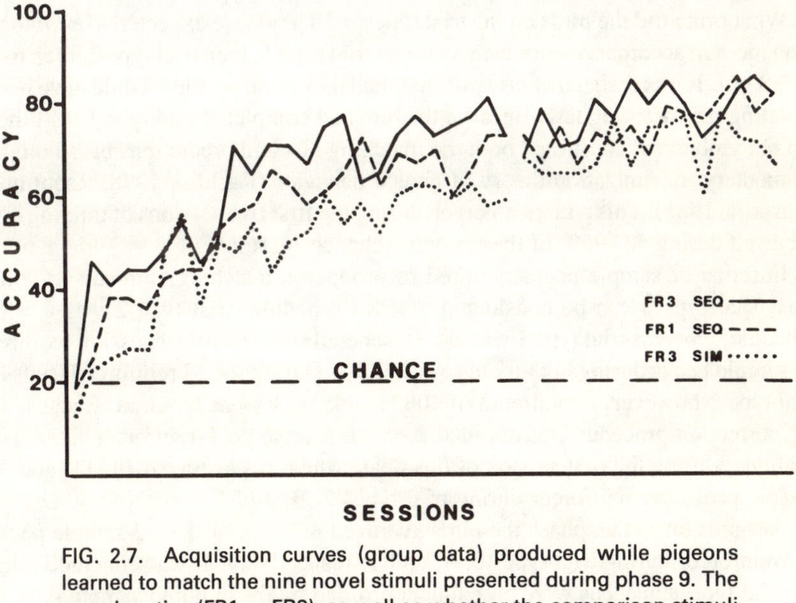

FIG. 2.7. Acquisition curves (group data) produced while pigeons learned to match the nine novel stimuli presented during phase 9. The sample ratios (FR1 or FR3), as well as whether the comparison stimuli were (SEQ)uentially or (SIM)ultaneously presented, is indicated as shown.

FIG. 2.8. The stimuli employed during phases 10 and 11.

The curves began to overlap during the last 15 session of the phase predominantly because of matching improvement during trial types 2 and 3.

What order did the birds do the trial types in? They were expected to emit their sequences in accordance with their success: trial type 1, then trial type 2, then trial type 3 last. It was realized afterwards that, had they done so, they would have been violating the matching law. Because the birds all completed trial type 1 last, they did not violate but confirmed both the matching law and predictions based on the momentary maximization theory of choice behavior (Rachlin, 1978; Rachlin & Krasnoff, 1983). This pattern emerged during the first five sessions of training and occurred during 53–95% of the sessions (chance = 16.6%).

In terms of sample pecks required to produce a matching comparison stimulus, trial type 3 can be considered an FR3 schedule, trial type 2 was a VR3 schedule, whereas trial type 1 was a VR9 schedule. An *errorless* session required 36 sample pecks during both trial types 2 and 3. During the 12 reinforced trials of trial type 1 however, a minimum of 108 sample pecks was required. Because of the correction procedure, errors increased the sample peck-reinforcer ratio. For example, during the first session of this phase, the birds averaged 18, 11, and 36 sample pecks per reinforcer during trial types 2, 3, and 1, respectively. During the last session of this phase the birds averaged 4.3, 3.6, and 10.3 sample pecks per reinforcer during trial types 2, 3, and 1, respectively. It was concluded that the trial type sequences were probably controlled by the required sample peck to reinforcer ratio rather than by the accuracy rates produced with the individual trial types.

Because differential accuracy levels had been obtained during phase 9, it seemed reasonable at this point to split the birds into two groups, and to train one group exclusively with trial type 2, the other group with trial type 3. The stimuli sets employed during the last two phases of the experiment are shown in Fig. 2.8.

During phase 10, birds 2 and 3 were trained to match the 5 novel stimuli of stimuli set 7 (2, 3, 5, 6, 11), whereas birds 1 and 5 were trained with stimuli set 8 (4, 7, 8, 9, 10). Birds 1 and 3 were trained on simultaneous-matching trials. The third sample peck produced all 5 comparison stimuli. Pigeons 2 and 5 were trained on sequential-matching trials. Each sample peck produced another comparison stimulus.

After 30 sessions of training, we again introduced a multiple matching condition, in phase 11. During half the reinforced trials, the familiar set of stimuli that the birds had learned to match during phase 10 was replaced with the 5 stimuli of

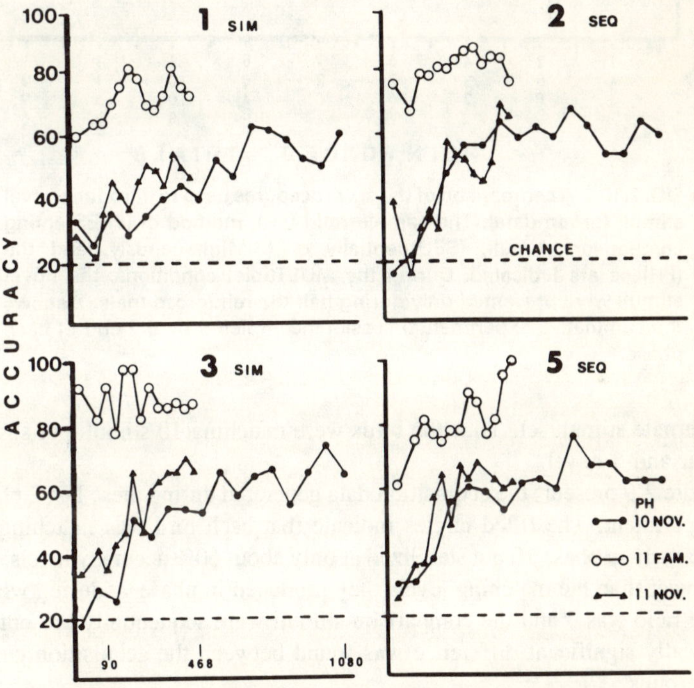

FIG. 2.9. Accuracy levels generated during phases 10 and 11. The number above each panel represents the bird. SIM or SEQ beside the number indicates whether the comparison stimuli were (SIM)ultaneously or (SEQ)uentially presented. Similar to Fig. 2.5, filled circles represent acquisition data generated while pigeons learned to match the five novel stimuli during phase 10. Open circles show matching levels produced with these now (FAM)iliar stimuli during phase 11. Filled triangles show the acquisition curves generated while pigeons learned to match the five novel stimuli during phase 11.

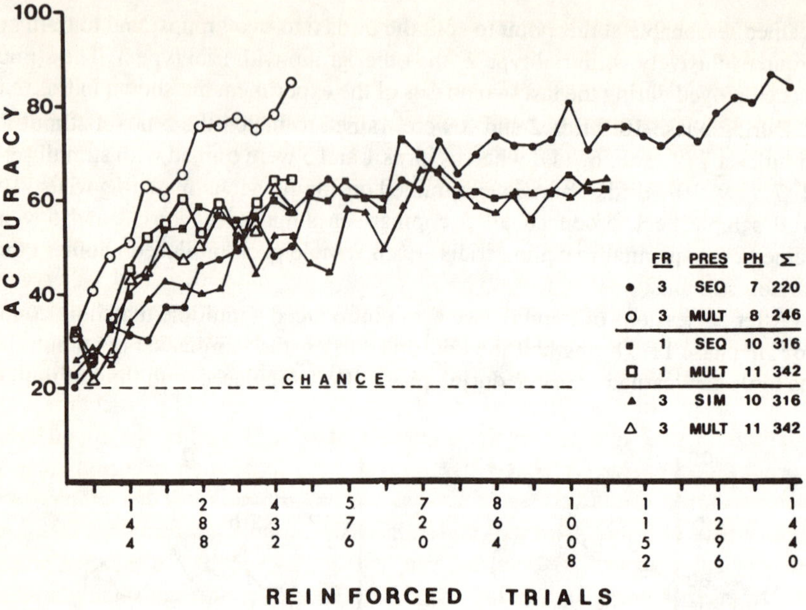

FIG. 2.10. A comparison of the six procedures used to introduce novel stimuli (group data). The sample ratio (FR), method of (PRES)enting comparison stimuli, (SEQ)uentially or (SIM)ultaneously, and the (PH)ase are indicated. During the MULT(iple) conditions, the novel stimuli were presented only during half the reinforced trials. Σ shows the cumulative experimental sessions completed by the end of each phase.

the alternate stimuli set. Thus the birds were matching 10 stimuli per session, 5 familiar and 5 novel.

Figure 2.9 presents the acquisition data generated during these last 2 phases of the experiment. The filled circles indicate that each bird was matching above chance early in phase 10 but stabilized at only about 60% accuracy levels; that is, 20% lower than the matching levels they produced in phase 7, during which the sample ratio was 3 and the comparison stimuli were sequentially presented. No statistically significant difference was found between the acquisition curves of the 2 groups.

Introducing the novel stimuli in phase 11 had little effect on the matching accuracies produced during the trials with familiar stimuli (open circles). Although the birds produced further improvement during the trials that presented the same stimuli employed in phase 10, reliable accuracy differences between the two groups did not emerge. During the acquisition trials with the novel stimuli, there was no statistically significant difference between the simultaneous and sequential groups. Although three of the birds produced some evidence of facilitated acquisition, there were no statistically significant differences between the acquisition curves of phases 10 and 11.

The lack of differences between the simultaneous and sequential matching groups was, to say the least, unexpected. A closer analysis of the data revealed a factor that may help explain the results. It is possible that birds 2 and 3 were not functionally on a schedule that sequentially presented comparison stimuli during all trials. At trial onset, the birds often emitted a burst of sample pecks that resulted in the presentation of three or four comparison stimuli. In other words, they often converted the trials into simultaneous discriminations and, as a result, produced matching levels comparable to those levels produced by the other birds. However, the results obtained during phase 9, presented in Fig. 2.7, strongly suggest that sequential presentation of comparison stimuli produces superior acquisition performance relative to simultaneous MTS procedures.

The last figure summarizes the acquisition data generated by the six different procedures that were employed. By far, the best acquisition occurred early in training during the multiple-matching conditions of phase 8 that employed a sample ratio of 3 and presented the comparison stimuli sequentially. This procedure yielded acquisition data that was statistically superior to the data generated by each of the other training schedules. There were no statistically significant differences among any of the other five training procedures, although the other two multiple-matching schedules did produce some trends towards facilitated matching acquisition.

Discussion

It appears that the three variables manipulated—sample ratio, sequential or simultaneous presentation of comparison stimuli, and employment of the multiple-matching condition—worked together to produce the superior transfer results that were observed. None of the three variables alone produced comparable acquisition data.

The multiple matching-to-sample procedure with a sample ratio of 3 and sequential presentation of the comparison stimuli produces facilitated matching performance for several reasons. First, higher sample ratios and longer sample durations have both been demonstrated to produce more rapid acquisition (Maki & Leith, 1973; Sacks, Kamil, & Mack, 1972; Zentall & Hogan, 1978). Second, presenting comparison stimuli sequentially reduces two sources of errors that occur when the comparison stimuli are presented simultaneously. Both key and comparison-stimulus bias from previous trials are lessened by not requiring all keys to be lighted nor all comparison stimuli to be present on every trial.

Finally, the multiple-matching condition probably reduces the influences of the sample and comparison stimuli from earlier trials as outlined in both Roitblat's (1983, 1984) drift theory, and Grant and Roberts (1976) competition decay model of proactive interference effects. Because familiar and novel stimuli sets are not mixed, two consecutive trials seldom present the same sample and comparison stimuli, which reduces presample and prechoice bias. Further, the multiple-matching condition provided 10 different samples per session compared to only 5 in the simple-matching condition. Lombardi, Fachinelli, and Delius (1984), and others have suggested that a larger number of exemplars will facilitate acquisition of a matching concept, by reducing the likelihood of stimulus control by only sample-specific rules or configuration discriminations (see Delius & Lombardi, this volume).

General Discussion

Several papers have suggested that the absolute properties of stimuli gain control first, followed by the emergence of multiple rules, i.e., two-component chains (Carter & Werner, 1978). If the situation is simple enough only to reinforce chained behavior, the relational factors among the stimuli (for example, whether they are the same or different) as a source of control will remain weak. Lombardi et al. (1984) suggested that birds are not purists. All three sources of control, configuration discriminations, two component chains, and a single rule or concept, are simultaneously able to influence the birds' behavior. Even if a single rule comes to dominate, the absolute properties of the stimuli probably still exert some influence due to both reinforcement history and proactive interference.

These factors support the position that the three-key MTS and OFS procedures may have limited applicability for the study of concept learning. One basis for this conviction are the results from a recent study conducted by Pisacreta, Potter, and Lefave (1984). In this study, the attempt was made to provide the easiest conditions possible under which matching transfer to novel stimuli might occur. Three pigeons were trained on the 3-key simultaneous MTS procedure. The sample was a 12-millimeter (mm) circle, cross, or triangle. Thirty pecks to the sample were required before the comparison stimuli were presented, and the sample was present when the bird made its choice. The pigeons required 105 sessions to produce 80% accuracy levels. In phase 2, the size of the stimuli was merely reduced to 8 mm. The accuracy levels decreased to near chance and required 65 sessions to recover. When the stimuli was reduced to 5mm in size during phase 3, the accuracy levels again fell, and the birds required 80 sessions of training to stabilize at 70–75% matching levels. What is to be concluded from the data? That the birds only learned two component chains involving 12-mm forms and then had to learn additional chains for 8-mm and 5-mm stimuli? How can one account for the disparity between their poor performances in this experiment and some of the facilitated matching transfer data reported elsewhere in this volume? Perhaps the limitations lie more in the present procedures than in the subjects.

Few people would claim that the pigeon's conceptual abilities will turn out to be extensive. Several reports, however, suggest that conceptual behavior is more likely to occur if the procedure is complex enough to make it necessary for a bird to learn a rule to produce reinforcers reliably. The typical MTS procedure trains the bird with two or three stimuli and then tests for facilitated transfer of matching behavior to one or two novel samples. When the bird fails to produce facilitated transfer, it is concluded that a matching concept is beyond the bird's ability. If three-key discriminations or two-component chains reliably produce grain, why should the bird be under the control of a generalized matching rule (see Pisacreta, 1982)? Barry Schwartz (1982) demonstrated that even humans will have difficulty learning a rule if repetition of a single response chain is reliably reinforced.

The results of this study lend support to the suggestion made by Zentall and his associates, i.e., efforts should be made to discover additional factors and procedures designed to facilitate transfer in conditional-discrimination paradigms. Based on recent studies (e.g., Herrnstein, 1985; Roitblat, Bever, & Terrace, 1984), procedures designed to produce concept learning in animals will be more likely to succeed if they: (1) employ a large number of stimuli to reduce proactive interference effects and make acquiring a rule more economical than learning dozens of chains or discriminations; (2) present complex stimuli—drawings and photographs—instead of hues or simple forms; (3) provide negative matching conditions during some of the trials; (4) present multiple-response locations to reduce success by chance and learning of only simple discriminations; (5) simultaneously introduce several novel matching problems; (6) use a correction procedure to reduce position preferences; and (7) require the animal to learn a rule. In effect, the procedure must *instruct* the animal to learn a rule.

Finally, the present work demonstrated a procedure that employs complex stimuli with complex situations and found that the birds showed, to some degree, evidence of rule-governed behavior, i.e., a same–different concept. Furthermore, the data support the position that an animal will learn a rudimentary concept if the complexity of the situation is structured to produce it (Herrnstein & Loveland, 1964).

Beyond our assessment of the pigeon's cognitive abilities, procedures that reliably produce evidence of rule-governed behavior in animals should help bridge the gap between psychology and ethology (see Shettleworth, 1983), as well as prove relevant to both education theory and the efforts to develop artificial intelligence.

ACKNOWLEDGMENTS

I would like to thank Lauraine Pisacreta for several editorial contributions, and David E. Gough, Richard Cross, Cathy Potter, William Schultz, Glenn Young, and James Kramer for their assistance in data collection and discussions.

REFERENCES

Carter, D. E., & Werner, T. J. (1978). Complex learning and information processing by pigeons: A critical analysis. *Journal of the Experimental Analysis of Behavior, 29,* 565–601.

Cerella, J. (1979). Visual classes and natural categories in the pigeon. *Journal of Experimental Psychology: Human Perception and Performance, 55,* 68–77.

Cumming, W. W., & Berryman, R. (1961). Some data on matching behavior in the pigeon. *Journal of the Experimental Analysis of Behavior, 4,* 281–284.

D'Amato, M. R., & Salmon, D. P. (1984). Cognitive processes in Cebus Monkeys. In H. L. Roitblat, T. G. Bever, & H. S. Terrace (Eds.), *Animal cognition* (pp. 149–168). Hillsdale, NJ: Lawrence Erlbaum Associates.

Delius, J. D., & Nowak, B. (1982). Visual symmetry recognition by pigeons. *Psychological Research, 44,* 199–212.

Farthing, G. W., & Opuda, M. J. (1974). Transfer of matching-to-sample in pigeons. *Journal of the Experimental Analysis of Behavior, 21,* 199–213.

Fujita, K. (1983). Acquisition and transfer of a higher order conditional discrimination performance in the Japanese monkey. *Japanese Psychological Research, 25,* 1–8.

Grant, D. S., & Roberts, W. A. (1976). Sources of retroactive inhibition in pigeon short-term memory. *Journal of Experimental Psychology: Animal Behavior Processes, 2,* 1–16.

Herrnstein, R. J. (1985). Riddles of natural categorization. *Philosophical Transcripts of the Royal Society of London, 308,* 129–144.

Herrnstein, R. J., & Loveland, D. H. (1964). Complex visual concept in the pigeon. *Science, 146,* 549–551.

Holmes, P. W. (1979). Transfer of matching performance in pigeons. *Journal of the Experimental Analysis of Behavior, 31,* 103–114.

Honig, W. K. (1978). Studies of working memory in the pigeon. In S. H. Hulse, H. Fowler, & W. H. Honig (Eds.), *Cognitive processes in animal behavior* (pp. 211–248). Hillsdale, NJ: Lawrence Erlbaum Associates.

Lea, S. G., & Ryan, M. E. (1983). Feature analysis of pigeon's acquisition of concept discrimination. In M. L. Commons, R. J. Herrnstein, & A. R. Wagner (Eds.), *Quantitative analysis of behavior, Vol. 4, discrimination processes* (pp. 239–253). Cambridge, MA: Ballinger.

Lombardi, C. M., Fachinelli, C. C., & Delius, J. D. (1984). Oddity of visual patterns conceptualized by pigeons. *Animal Learning and Behavior, 12,* 2–6.

Maki, W. S., & Leith, C. R. (1973). Shared attention in pigeons. *Journal of the Experimental Analysis of Behavior, 19,* 345–349.

Pisacreta, R. (1982). Preferences among stimulus matches in the pigeon. *Journal of the Experimental Analysis of Behavior, 38,* 191–199.

Pisacreta, R., Lefave, P., Lesneski, T., & Potter, C. (1985). Transfer of oddity learning in the pigeon. *Animal Learning and Behavior, 13, (4),* 403–414.

Pisacreta, R., Potter, C., & Lefave, P. (1984). Matching of varying-size form stimuli in the pigeon. *Bulletin of the Psychonomic Society, 22*(6), 591–593.

Pisacreta, R., Redwood, T., & Witt, K. (1984). Transfer of matching to figure samples in the pigeon. *Journal of the Experimental Analysis of Behavior, 42,* 223–237.

Pisacreta, R., & Witt, K. (1983). Same–different discriminations in the pigeon. *Bulletin of the Psychonomic Society, 21*(5), 411–414.

Premack, D. (1978). On the abstractness of human concepts: Why it would be difficult to talk to a pigeon. In S. H. Hulse, H. Fowler, & W. K. Honig (Eds.), *Cognitive processes in animal behavior* (pp. 423–451). Hillsdale, NJ: Lawrence Erlbaum Associates.

Rachlin, H. (1978). A molar theory of reinforcement schedules. *Journal of the Experimental Analysis of Behavior, 30,* 345–360.

Rachlin, H., & Krasnoff, J. (1983). Eating and drinking: An economic analysis. *Journal of the Experimental Analysis of Behavior, 39,* 385–404.

Roitblat, H. L. (1983). Pigeon's working memory: Models for delayed matching-to-sample performance. In M. L. Commons, R. J. Herrnstein, & A. R. Wagner (Eds.), *Quantitative analyses of behavior, Vol. 4, Discrimination processes* (pp. 161–181). Cambridge, MA: Ballinger.

Roitblat, H. L. (1984). Representations in pigeon working memory. In H. L. Roitblat, T. G. Bever, & H. S. Terrace (Eds.), *Animal cognition* (pp. 79–97). Hillsdale, NJ: Lawrence Erlbaum Associates.

Roitblat, H. L., Bever, T. G., Terrace, H.S. (Eds.). (1984). *Animal cognition.* Hillsdale, NJ: Lawrence Erlbaum Associates.

Sacks, R. A., Kamil, A. C., & Mack, R. (1972). The effects of fixed ratio sample requirements on matching-to-sample in the pigeon. *Psychonomic Science, 26,* 291–293.

Schwartz, B. (1982). Reinforcement-induced behavioral stereotypy: How not to teach people to discover rules. *Journal of Experimental Psychology: General, 111,* 23–59.

Shettleworth, S. J. (1983). Memory in food hoarding birds. *Scientific American, 248* (3), 102–110.

Urcuioli, P. J., & Nevin, J. A. (1975). Transfer of hue matching in pigeons. *Journal of the Experimental Analysis of Behavior, 24,* 149–155.

Wright, A. A., Santiago, H. C., Urcuioli, P. J., & Sands, S. F. (1983). Monkey and pigeon acquisition of same/different concept using pictorial stimuli. In M. L. Commons, R. J. Herrnstein, & A. R. Wagner (Eds.), *Quantitative analyses of behavior, Vol. 4, Discrimination processes* (pp. 295–317). Cambridge, MA: Ballinger.

Zentall, T. R., Edwards, C. A., & Hogan, D. E. (1983). Pigeon's use of identity. In M. L. Commons, R. J. Herrnstein, & A. R. Wagner (Eds.), *Quantitative analyses of behavior, Vol. 4, Discrimination processes* (pp. 273–293). Cambridge, MA: Ballinger.

Zentall, T. R., Edwards, C. A., Moore, B. S., & Hogan, D. E. (1981). Identity: The basis for both matching and oddity learning in pigeons. *Journal of Experimental Psychology: Animal Behavior Processes, 7,* 70–86.

Zentall, T. R., & Hogan, D. E. (1978). Same/different concept learning in the pigeon: The effect of negative instances and prior adaption to transfer stimuli. *Journal of the Experimental Analysis of Behavior, 30,* 177–186.

Zentall, T. R., Hogan, D. E., & Edwards, C. A. (1980). Oddity learning in the pigeon: Effect of negative instances, correction, and number of incorrect alternatives. *Animal Learning and Behavior, 8 (4),* 621–629.

3 Size Invariance in Visual Pattern Recognition by Pigeons

Celia M. Lombardi
Juan D. Delius
Experimentelle Tierpsychologie, Psychologisches Institut, Federal Republic of Germany

INTRODUCTION

The visual recognition of objects and scenes plays a crucial role in the life of most higher animals, including humans. Many field studies document that the identification of items such as the home patch, the breeding site, the neighbor, the mate, and the offspring is essential for the survival and reproduction of many animals. Furthermore, the recognition of certain collections of scenes and objects as belonging to separate categories, as for example habitats, nesting sites, edibles, conspecifics, relations, or predators, has been demonstrated to be equally vital for higher animals. Indeed, it is already known thanks mainly to the efforts of the Harvard school that, besides humans, at least some vertebrates, and particularly pigeons, are capable of comparable recognition performances in controlled laboratory experiments. It is likely that this competence is not restricted to ecologically relevant stimuli, but that it also extends to at least some less natural stimuli (Cerella, 1982; Delius & Nowak, 1982; Herrnstein, 1985; Lea, 1984).

Little is known about the information-processing operations that underlie these capacities but they clearly are complex. An analysis might be easier if the processes leading to visual recognition could be broken down into components that the visual system has to deal with sequentially. There is one initial and fundamental problem with which most, if not all, central visual systems have to cope. Even when the relevant objects or scenes are static, as inanimate objects and scenes usually are, the information that the central visual system has to operate on, as a rule, is not. As a function of the orientation, the distance, the lighting, and so forth at the time of viewing, the successive retinal images corresponding to a given scene or object vary (Fig. 3.1). If the appearance of the object or scene additionally changes by itself, as is often the case if it is alive, then the problem is compounded. Parenthetically it is noted here that the time course of the change, most often in the form of patterned movement, may in itself be an important cue for recognition, a subject that Emmerton (this volume) pursues elsewhere in this book.

Processes designed to cope with the inevitable variability of the retinal images of even one and the same scene or object are thus essential for any visual system. In the human-research literature these have been reported as perceptual invariance phenomena and are seen as contributing to what is termed the subjective constancy of objects and scenes. In a series of experiments that used the matching-to-sample, or conversely the oddity-from-sample procedure, the extent to which the recognition of two-dimensional shapes in pigeons is independent of various kinds of

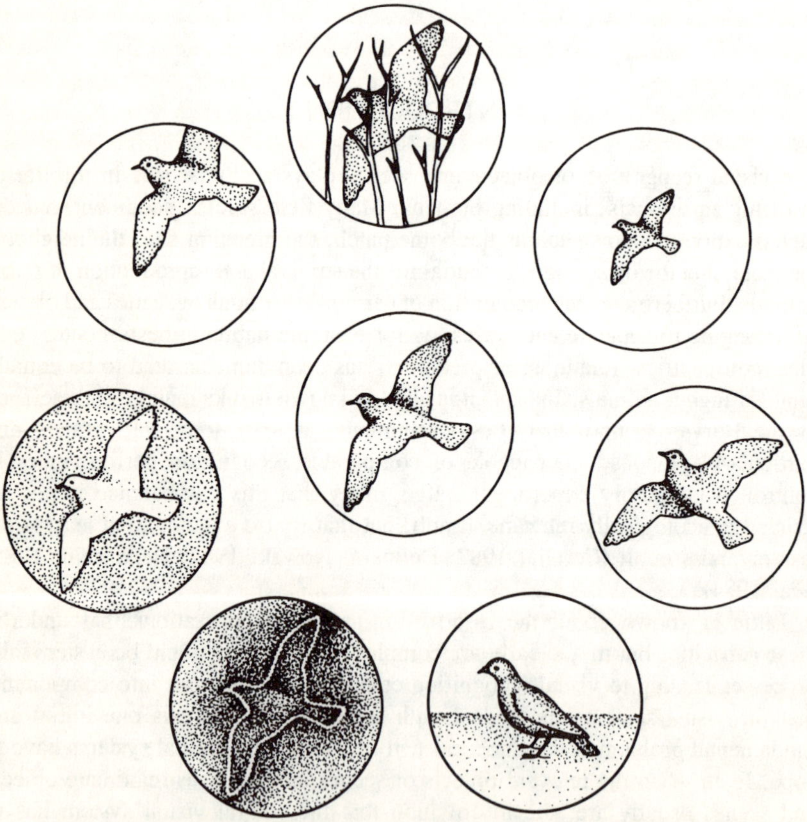

FIG. 3.1. Illustration of some of the situations where invariance operations are required to recognize an object.

stimulus transformations has been examined. Elsewhere, the present researchers have reported on orientation invariance (Delius, 1986; Delius & Hollard, 1987; Hollard & Delius, 1982; Lombardi, in press). A further paper will deal with contrast, outline, and color invariance of pattern recognition (Lombardi & Delius, in preparation).

Results relating to the size invariance of shape recognition are presented here. Again, most visual systems, whether natural or artificial, have to cope with the fact that the size of the image of an object on the phototransductive surface (retina or otherwise) varies with the viewing distance. The ability to recognize images as identically shaped even when their sizes are different is a desirable property of any efficient image-processing system. In humans this capacity is closely associated with the phenomena of size constancy (a given object is subjectively perceived as being of a constant size when looked at from different distances) and of distance perception (when yielding a small retinal image a given object is interpreted as being further away than when producing a large image: successive views of an approaching car, for example). At this stage no inquiry will be made as to whether that also applies to pigeons.

In humans, again, the size-invariance competence extends to situations where viewing distance is in fact held constant and the actual objects or scenes vary in scale. This is ecologically meaningful as in man-made environments equivalent objects and scenes do frequently occur in different sizes (successive zoom takes in a film, for example), although in natural environments the range over which they do so often tends to be quite restricted (different sized oak leaves, for instance). For the sake of methodological simplicity, the assumption will be made that pigeons are the same as humans in this respect although one cannot be sure. Judging from available artificial-vision evidence (see references later), it appears almost inevitable that whatever size-invariance mechanism is implemented, it will serve indiscriminately in both the variable viewing distance/constant object size and the constant viewing distance/variable object size conditions.

Though to humans size invariance appears subjectively to be a trivially easy operation, it has proved to be an exceedingly demanding task for such artificial systems. Although considerable effort has gone into producing appropriate hardware and software, their performance is still rudimentary (e.g., Glünder, 1986; Hord, 1982; Reitböck & Altmann, 1984), not least because of the discrepancy between the vast computing power demanded and that economically viable at present. Incidentally, each kind of invariance (orientation, size, etc.) appears to require a rather specific transformation of the pictorial information. Pigeons, which have a brain much smaller than humans, might accordingly be expected to be short of processing capacity even though, as fast-moving animals, size-invariance capacities are likely to be essential for their fitness.

It needs stressing that in humans the competence for size-compensated object or scene identification is spontaneous in the sense that they do not need repeated experience with a given object to apply size invariance to it. Indeed, it is a question of whether the ontogenetic development of this ability depends on specific learning or whether it is largely innate. A capacity for spontaneous size invariance with novel stimuli is also to be expected of pigeons on general functional grounds.

Some evidence already indicates that pigeons continue to distinguish two different visual shapes in a simultaneous conditioning paradigm when the sizes of these shapes are altered during test trials (Towe, 1954). In several perceptual categorization experiments, in which pigeons had to recognize different kinds of complex shapes, they seemed to do so regardless of the size of the targets (e.g., Cerella, 1979; Delius & Nowak, 1982; Herrnstein, Loveland, & Cable, 1976; Lea & Ryan, 1983; Siegel & Honig, 1970; see also Ryan & Lea, this volume, for a more special case concerning chickens). In a matching-to-sample experiment, akin to those presented here, Pisacreta, Potter, and Lefave (1984) found that pigeons trained on shapes that were all the same size continued to show matching behavior when, at a certain point, the size of the sample and comparison shapes was suddenly made to differ. Because the birds had extensive previous experience with the shapes in a standard size, Pisacreta et al. considered that stimulus generalization might have been responsible for the transfer obtained. It is debatable though whether this explanation is satisfactory. The authors did not measure in any detail the generalization gradients associated with the training stimuli they used. Conventional generalization gradient experiments show that pigeons continue to respond, although making fewer responses, to a shape altered in size after their having been conditioned to the same shape in a particular size (e.g., Jenkins, Pascal, & Walker, 1958; Wildemann & Holland, 1973). These experiments, however, do not by themselves imply size invariance, as they do not examine the degree to which shape discrimination is retained across size variations. The extent to which the phenomenon that underlies generalization gradients can nevertheless be invoked to account for size invariance of pattern recognition needs careful scrutiny. The issue is discussed later.

SIZE INVARIANCE WITH ARBITRARY ODD SHAPES

In this experiment homing pigeons (Columbia livia) of local origin were used. They had extensive previous experience with an oddity-from-sample task involving many different three-stimulus sets. They had been shown to be capable of solving the oddity problem, at least partially, on a relational basis (Lombardi, Fachinelli, & Delius, 1984) and had since served in invariance experiments involving variables other than size (Lombardi, in press).

A Skinner box with three keys was employed. Each key was provided with a shutter. An automatic projector, taking special slides with coding perforations read by photocells, projected the stimuli on the keys. The apparatus was controlled by a computer that also recorded the subjects' responses (Fig. 3.2). Daily

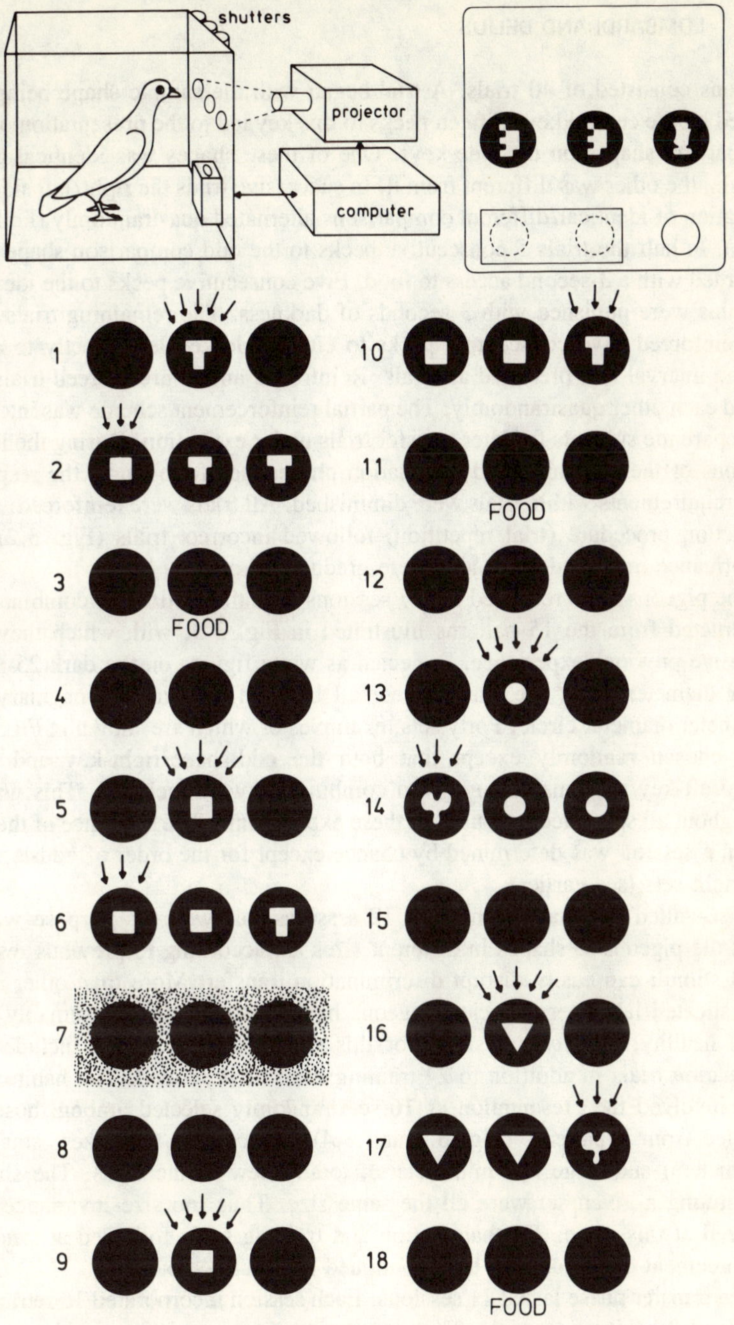

FIG. 3.2. Top: Apparatus and slides employed. Bottom: Illustration of the procedure used. Lines 1 to 4 correct trial, food reward; 5 to 8 incorrect trial, punished with darkness; 9 to 12 correction trial, stimulus configuration repeated; 13 to 15 nonreinforced correct trial, 16 to 18 correct trial, food reinforcement.

sessions consisted of 40 trials. A trial began with the sample shape being displayed on the central key. Fifteen pecks to this key led to the presentation of two comparison shapes on the side keys. One of these shapes was identical to the sample, the other was different from it. In successive trials the right/left side key allocation of identical/different comparisons alternated quasirandomly (Fellows, 1967). In half the trials 5 consecutive pecks to the odd comparison shape were rewarded with a 3-second access to food. Five consecutive pecks to the identical stimulus were punished with 3 seconds of darkness. The remaining trials were not reinforced; five consecutive pecks to either side key led directly to an 8-second interval that preceded all trials. Reinforced and nonreinforced trials followed each other quasirandomly. The partial reinforcement scheme was intended to prepare the subjects for later transfer trials under extinction. During the initial sessions of the retraining and habituation phases (see following), the response ratio requirements within trials were diminished. All trials were reinforced, and a correction procedure (trial repetition) followed incorrect trials (Fig. 3.2). As performance improved these aids were gradually dropped.

The pigeons were retrained for 17 sessions with three-stimulus combinations constructed from the 15 patterns illustrated in Fig. 3.3, with which they had extensive previous experience. Projected as white figures on the dark 25-millimeter diameter keys, the shapes were all barely fitting in an imaginary 10-millimeter diameter circle. Forty sets, examples of which are shown in Fig. 3.4, were chosen randomly except that both the odd-shape/right-key and odd-shape/left-key versions of any given combination were included. This was so throughout all set collections used in these experiments. The sequence of the sets within a session was determined by chance except for the order of odd-left and odd-right sets (see earlier).

A so-called habituation phase of 12 sessions followed. Its purpose was to adapt the pigeons to shapes in different sizes as, according to previous results, novel stimuli can easily disrupt discrimination transfer. More than other more domesticated laboratory species, pigeons have a strong, and presumably normally healthy, novelty aversion. For this reason these sessions included 16 habituation trials in addition to 24 training trials as described. The habituation trials involved the presentation of 16 sets randomly selected among those assembled from 5 patterns (Fig. 3.3 and 3.4) projected in two sizes, small (5 millimeters) and large (20 millimeters), totally new to the birds. The shapes constituting a given set were all the same size. Thus, no size invariance was required at this stage. The habituation and training trials followed at random; reinforcement conditions for both continued as described before.

The transfer phase lasted 11 sessions. Each session incorporated 16 retraining trials and 8 habituation trials. The sets used in a given session for these trials were randomly selected from among those used previously. The remaining 16 trials were transfer trials, involving for the first time sets with size differences between the samples and the comparison shapes. They were inserted randomly

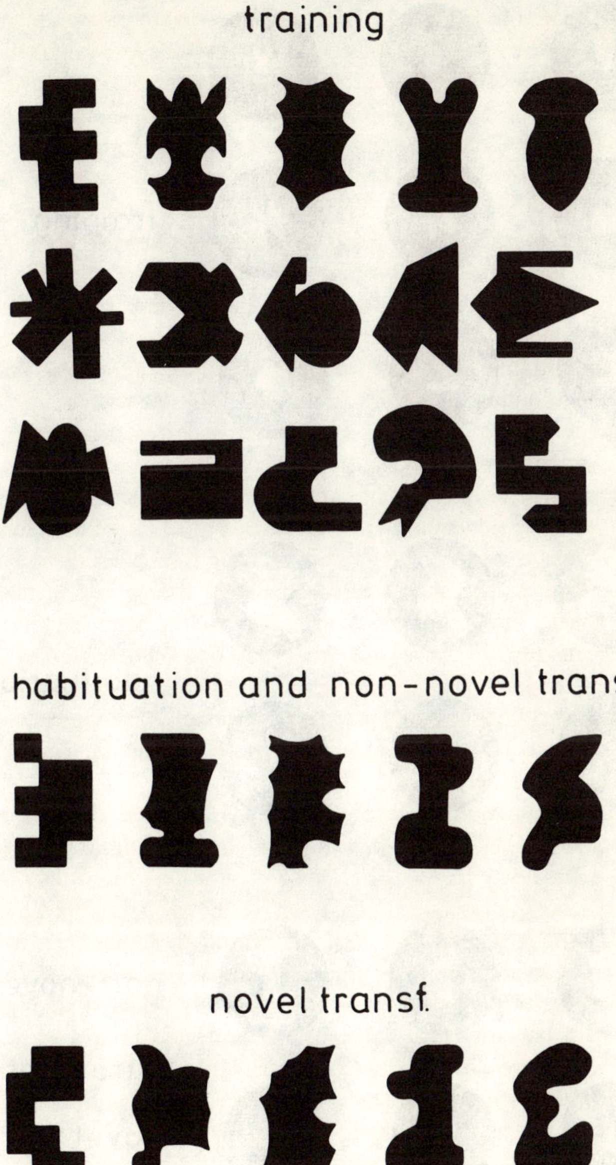

FIG. 3.3. Shapes used in the first experiment.

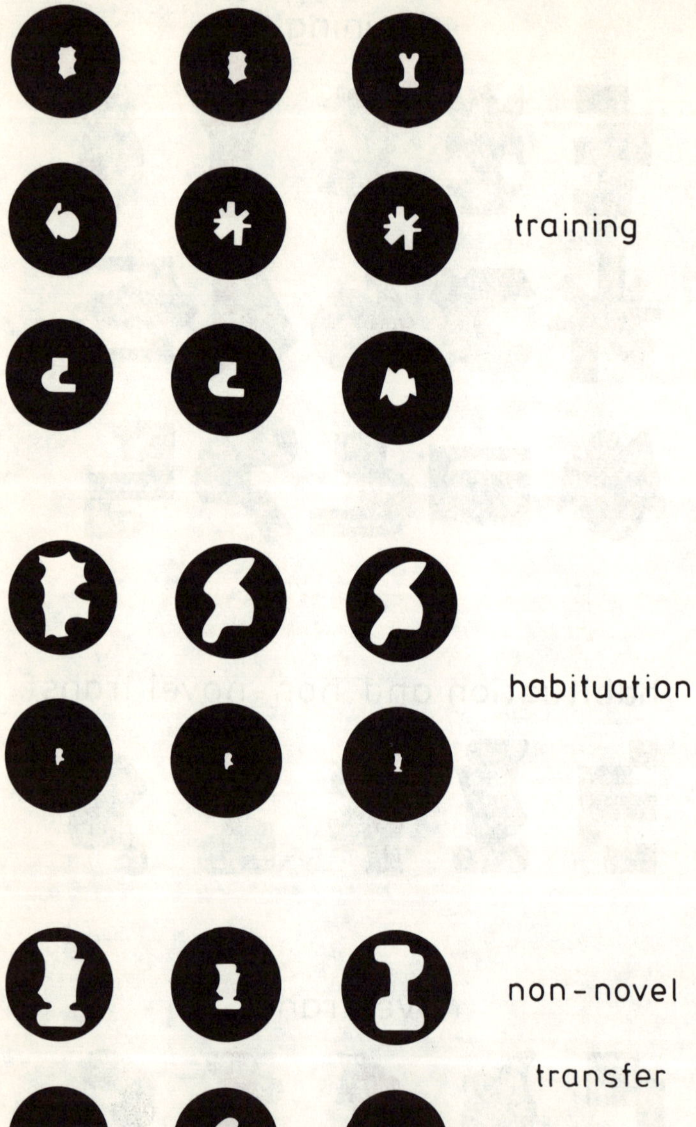

FIG. 3.4. Examples of sets used in the first experiment.

among the other types of trials, but so as to avoid the first three and last three trials of a session. Two collections of 40 sets constructed from two separate groups of patterns were presented (Fig. 3.3). One group of 5 patterns was the same used for composing the habituation sets, which the pigeons therefore knew in small and large size. The other 5 figures were novel in the sense that the pigeons had previously experienced them temporarily but exclusively in medium (10 millimeter) size and only in nonreinforced trials about 6 months earlier (Lombardi, Fachinelli, & Delius, 1984). The sample shapes of all transfer sets were medium size, but both comparison patterns of a given set were either small (5 millimeters) or large (20 millimeters; Fig. 3.4). Each three-stimulus combination existed in a small- and a large-comparison-shape version. Without regard to the size variants, each shape appeared equally often as a sample, a matching, and an odd stimulus.

The average percent-correct trials, computed separately for various components, are shown plotted session by session in Fig. 3.5. The pigeons' mean performance on the transfer trials involving novel shapes and incorporating size disparities is shown singly for sets with small and with large comparison shapes. It was well above-chance level except for the large comparison sets in one session. The small comparison sets generally yielded better results than the large comparison sets. As trials were only partially reinforced (see earlier), the mean performance on the individual novel transfer sets before any responses to them had been reinforced could be calculated separately (Fig. 3.5). They represent the true size-invariance tests of this experiment. The performance on transfer sets made from stimuli with which the birds already had reinforced experience was very closely similar and only minimally better. It is not shown.

The results, summarized as overall average percent-correct trials on various components during the last 11 sessions before the transfer phase and during the 11 sessions of the transfer phase, are depicted in Fig. 3.6. They emphasize that the pigeons continued to evince oddity-from-sample discrimination in the size-invariance tests proper (novel shapes, not reinforced). Binomial tests showed that each subject did so at above-chance level (all P's < 0.05 or less). However, the accuracy was better on the small than on the large comparison sets (Wilcoxon, $P < 0.01$). This trend applied generally to all transfer components. The performance on the small and the large habituation sets, though not involving size discrepancies, also differed in the same direction (Wilcoxon, $P < 0.01$). During the transfer phase the time between the onset of the comparison stimuli and the first response to one of them was recorded as reaction time for each trial. Figure 3.6 also displays the averages of these reaction times corresponding to some components. Generally, reaction times were longer when small, rather than large patterns, were involved. This difference was appreciable and significant during the nonreinforced size-invariance test trials with the novel shapes (Wilcoxon, $P < 0.05$). The implications of these findings are discussed later.

The main result of this first experiment is that pigeons were capable of

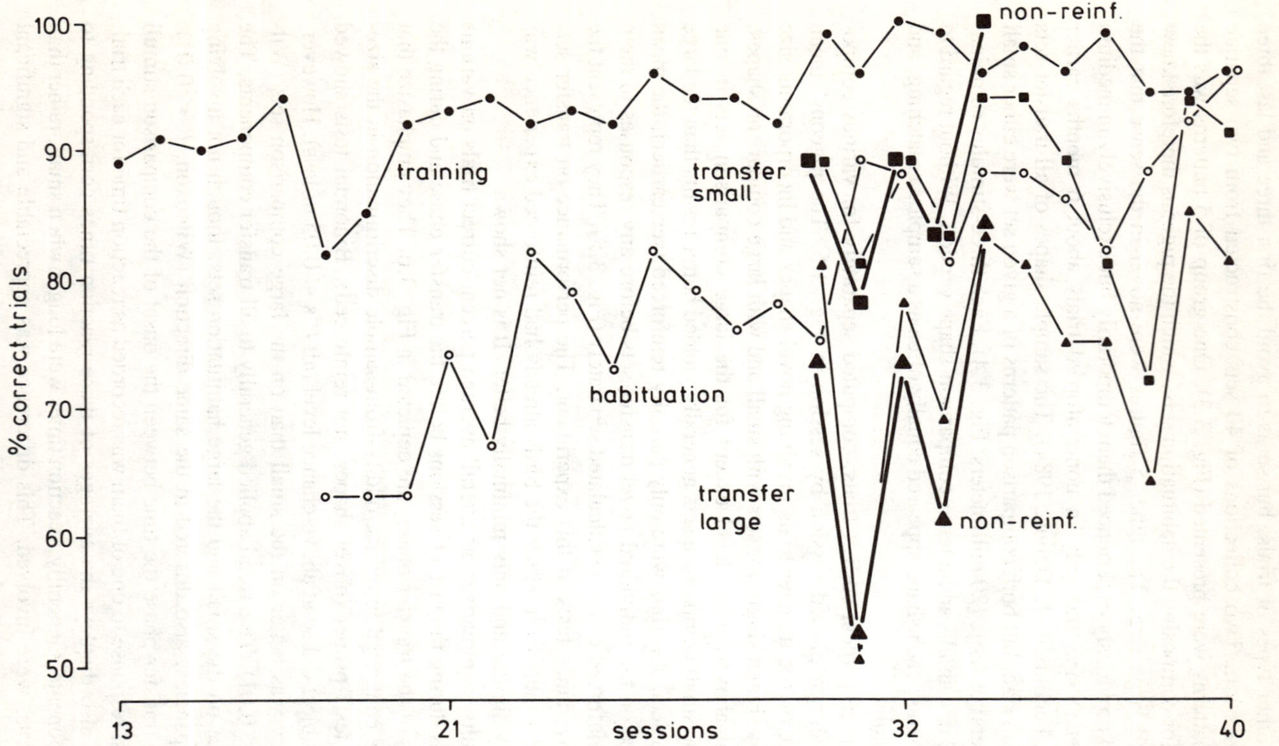

FIG. 3.5. Average performance of pigeons on various components during the first experiment. Initial 12 training sessions are not shown. Performance on training (filled circles), habituation (open circles), and transfer trials (novel shapes only, triangles: large comparison stimuli, squares: small comparison stimuli; large symbols, thick lines nonreinforced trials). For summary see Fig. 3.6.

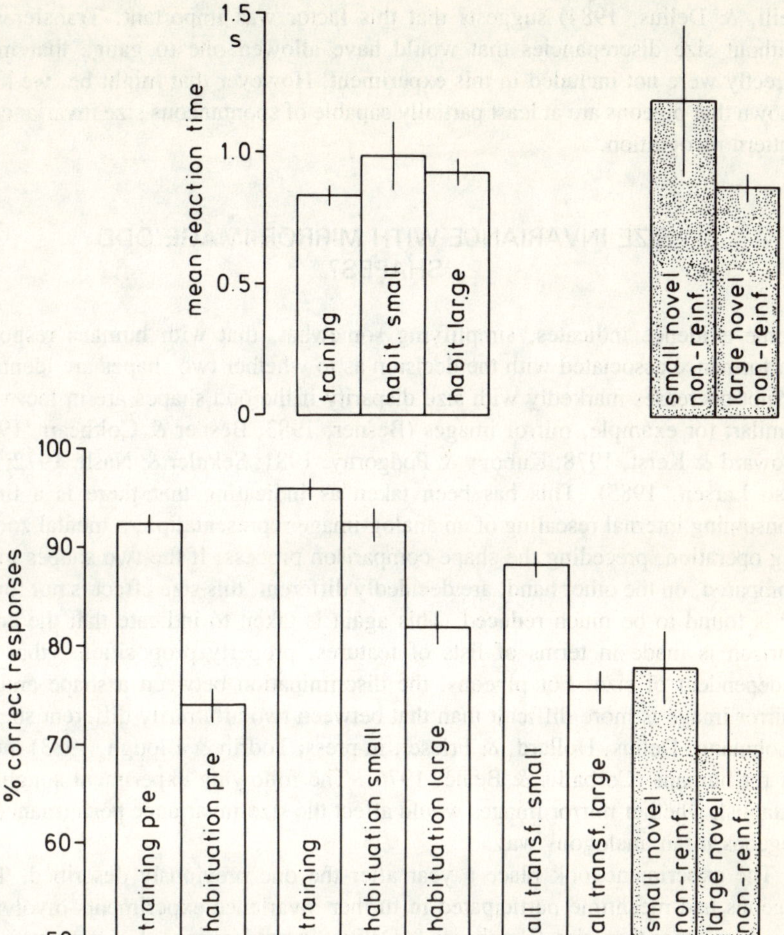

FIG. 3.6. Average accuracy and average speed both with standard deviations on the various components of the first experiment. Shaded: true invariance tests.

spontaneously recognizing at above-chance levels the same/different nature of visual patterns when the comparanda differed in linear size by a factor of two. It is important that they achieved this with shapes that had never been associated with reinforcement and that they had only been seen previously in one standard size. To what extent the less than perfect transfer performance was due to the size disparities or the imperfect transference of the relational same/different rule to the new stimulus material is somewhat uncertain. The lesser decrement observed in earlier oddity-transfer tests not involving size differences (Lombardi, Fachinelli, & Delius, 1984) suggests that this factor was important. Transfer sets without size discrepancies that would have allowed one to gauge that more directly were not included in this experiment. However that might be, we have shown that pigeons are at least partially capable of spontaneous size invariance in pattern recognition.

SIZE INVARIANCE WITH MIRROR-IMAGE ODD SHAPES?

Some evidence indicates, simplifying somewhat, that with humans response performance associated with the decision as to whether two shapes are identical or not decreases markedly with size disparity if the odd shapes are in fact very similar; for example, mirror images (Besner, 1983; Besner & Coltheart, 1976; Howard & Kerst, 1978; Kubovy & Podgorny, 1981; Sekuler & Nash, 1972; see also Larsen, 1985). This has been taken as indicating that there is a time-consuming internal rescaling of an analog-image representation, a mental zooming operation, preceding the shape-comparison process. If the two shapes to be compared, on the other hand, are decidedly different, this size effect is not found or is found to be much reduced. This again is taken to indicate that the comparison is made in terms of lists of features, property propositions, that are independent of size. For pigeons, the discrimination between a shape and its mirror image is more difficult than that between two arbitrarily different shapes (Lohmann, Delius, Hollard, & Friesel, in press; Todrin & Blough, 1983) much as for humans (Corballis & Beale, 1976). The following experiment sought to establish whether mirror images would affect the size-invariance performance of pigeons in an analogous way.

The experiment took place a year after the one previously described. The pigeons had meantime participated in further invariance experiments involving variables other than size (Lombardi & Delius, in preparation). For not precisely understood reasons the performance of 3 animals on the straight oddity-from-sample task, bare of any invariance component, had badly deteriorated, and they had meantime been dropped. The 6 remaining pigeons were retrained for 9 sessions with new collections of 40 randomly selected sets based on the 15 training patterns illustrated in Fig. 3.3. The odd comparison stimuli of these sets were still arbitrarily different from the sample stimuli. The habituation phase that followed lasted 25 sessions. Twenty-four trials of these sessions involved 24 habituation sets constructed from 4 figures. Each figure was used in 3 sizes, 5, 10, and 15 millimeters. The shapes in a given set were all the same size. The odd patterns were now always mirror images of the samples (Fig. 3.7).

The transfer phase lasted 10 sessions that consisted of 16 training, 8 habituation, and 16 transfer trials. Two collections of 40 transfer sets each were prepared from two different groups of 7 figures. Five patterns of each group were

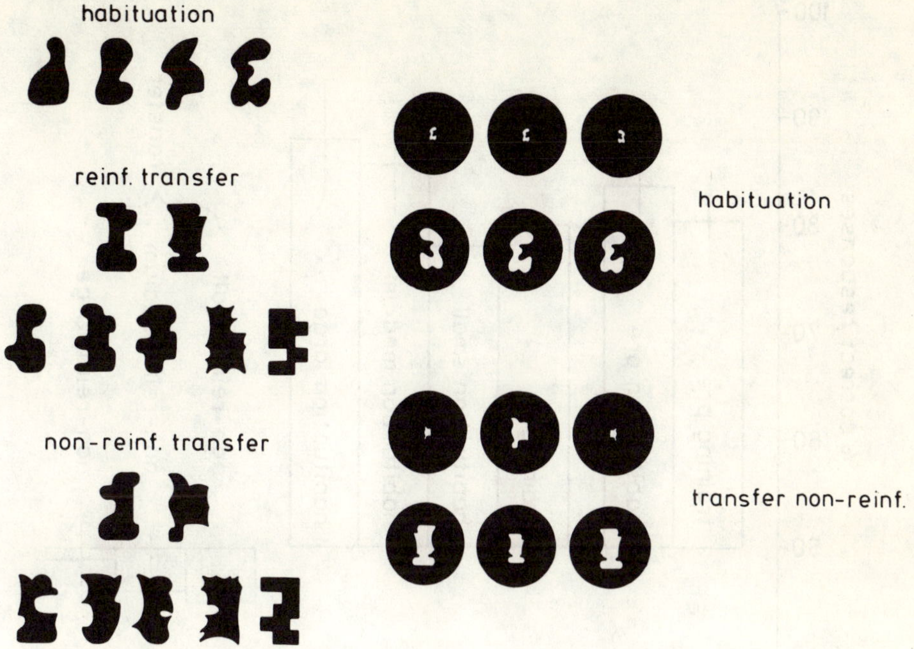

FIG. 3.7. Shapes and examples of sets used in the second experiment.

novel for the pigeons; the other two had already been used during the first experiment. All shapes occurred in 3 sizes. The sample patterns were always 10 millimeters in size, but the comparison patterns were both 5, 10, or 15 millimeters (Fig. 3.7). Each set was shown twice in the course of the transfer phase. One collection of transfer sets was consistently reinforced whereas those of the other collection, i.e., the proper test sets, were never reinforced. The idea of having one transfer collection reinforced was that the animals would not be able to associate novelty, which can be recognized by pigeons as a stimulus category of its own (Delius, 1985), with nonreinforcement.

The most striking fact is the pigeons' chance-level performance on the transfer sets. This is detailed in Fig. 3.8 for the nonreinforced sets, but it applied equally to the reinforced sets (49.5% correct trials overall). It is important that the transfer failure also affected the sets involving no size differences. This indicates that the difficulty not lie primarily in size invariance. Rather, the pigeons in this instance failed to transfer the basic same/different discrimination to the novel sets.

It could be suspected that the reason for the failure was the use of mirror-image odd shapes. The final performance on the habituation sets that also incorporated mirror images was, however, excellent. Furthermore, in other invariance

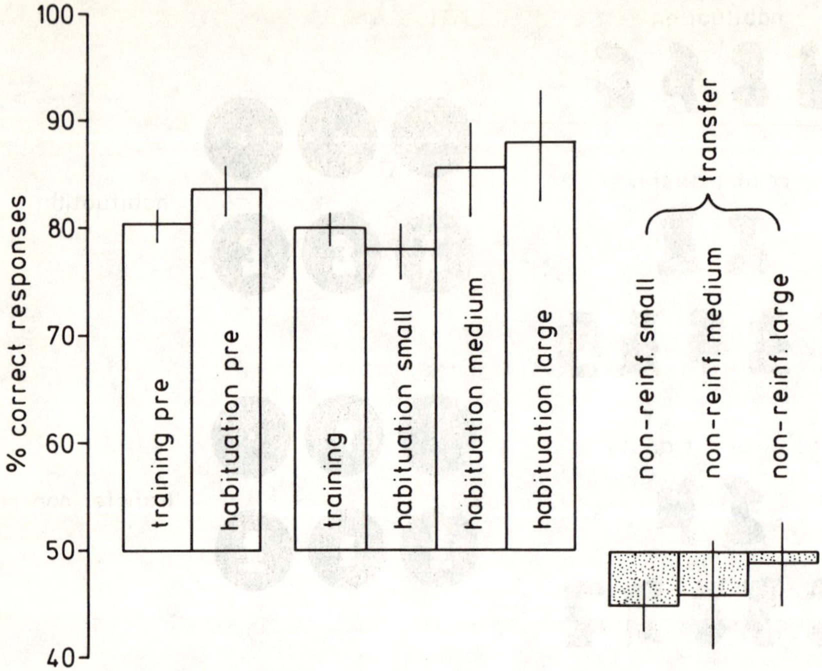

FIG. 3.8. Average accuracy with standard deviations on the various components of the second experiment. Shaded: critical transfer test.

experiments, mirror images did not impede good transfer performance (Delius & Hollard, 1987; Hollard & Delius, 1982). More likely, in retrospect, the failure was due to the fact that the experimenters disregarded their own advice about the necessity of many training exemplars (Lombardi, Delius, & Hollard, 1986; Lombardi, Fachinelli, & Delius, 1984) by using too few habituation sets, the only exemplars involving mirror images. This might have invited the pigeons to solve the oddity task in a nonconceptual way that prevented transfer to new stimulus sets, so that they could not apply their size-invariance capabilities. Whether the complexity of the size-invariance task contributed to a quitting attitude by the subjects is a matter of speculation. In any case, the experiment certainly needs to be replicated with a more propitious design before accepting this particular inability of pigeons as an established fact.

The reason for presenting this experiment, however, is that in one special way its totally negative results are very reassuring. They demonstrate that the positive outcome of the first experiment and other invariance experiments done by us with exactly the same technique cannot be due to an insidious, unintended, artifactual cue. If such existed it should also have been effective in the present experiment.

INFLUENCE OF RELATIVE AND ABSOLUTE SIZE

In experiments with humans, Bundesen and Larsen (1975; see also Bundesen, Larsen, & Farrel, 1981), among others, found that the performance associated with deciding whether two different-sized figures had the same shape depended on the size differences between the figures, but not on their absolute size. This result further supports the mental-zooming hypothesis sketched earlier. Because the design of the first experiment presented here did not allow inferences as to the interaction of relative and absolute size, a third experiment was conceived to investigate this question.

The same 6 pigeons that 3 months earlier had completed the previous experiment served as subjects again. In the interval they had been kept experimentally idle in an outside aviary. Retraining for no fewer than 65 sessions was necessary, because they had forgotten the basic oddity-from-sample task. Incidentally, this reflects an interesting procedural-memory weakness that contrasts with the excellent long-term visual item memory of pigeons (Delius, 1985; Vaughan & Greene, 1984). Twenty stimulus sets randomly chosen from among those used previously, plus 20 new sets randomly selected from those based on the customary training figures (Fig. 3.3) were employed.

During a habituation phase lasting 32 sessions, the animals were exposed to 2 collections of 40 sets picked at random from among those based on 2 separate groups of 5 novel patterns (Fig. 3.9). The figures appeared in 4 different sizes, 5, 10, 15, and 20 millimeters as projected on the pecking keys. All 3 stimuli of a given set were the same size (Fig. 3.9). The first habituation session included only 4 habituation trials, but their number was gradually increased and from the 18th session onwards all 40 trials of a session were habituation trials.

The transfer phase consisted of 20 sessions. Each included 16 training and 8 habituation trials. The 16 transfer trials involved sets made up from the same shapes that were used for habituation (Fig. 3.9). This departed from earlier routine and was intended to facilitate the pigeons' task in an otherwise complex transfer design. Differently from the previous experiments, the sample as well as the comparison patterns could appear in any of the 4 sizes. The size differences between sample and comparison shapes present in a given set could be graded from 0 to 3 (Table 3.1). One group of figures yielded the sets employed in nonreinforced trials (the true invariance tests) and the other, those used in reinforced trials. Two collections of 160 sets were randomly selected from those assembled with the 2 groups of figures. All size differences occurred equally often, but this meant that it was not possible to balance the occurrence of the different absolute sizes fully. The difference 1 for instance could arise in 6 different ways involving all sizes, but the difference 3 only in 2 ways involving only 2 sizes.

The average correct trials on the transfer sets with the four size differences were calculated separately for the reinforced and the nonreinforced set collec-

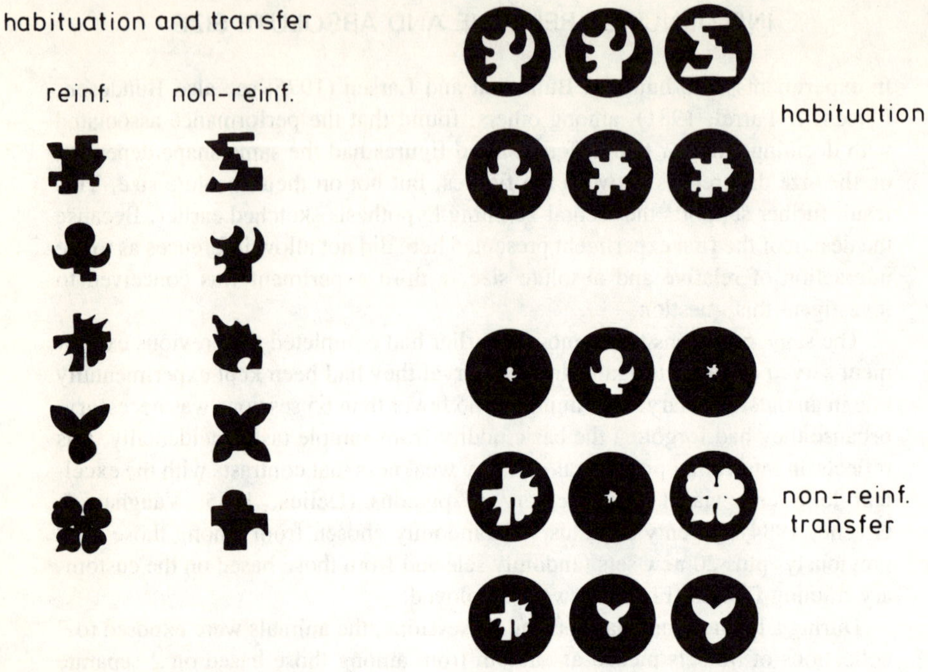

FIG. 3.9. Shapes and examples of sets used in the third experiment.

TABLE 3.1
Size Differences Between Sample and Comparison Shapes

		Size of Sample			
		5mm	10mm	15mm	20mm
Size of Comparisons	5mm	0	1	2	3
	10mm	1	0	1	2
	15mm	2	1	0	1
	20mm	3	2	1	0

tions (Fig. 3.10). When only the nonreinforced sets were considered, they yielded above-chance accuracy when the differences 0 and 1 were present (binomial tests, each pigeon, each difference $P < 0.05$ or less, except one subject in the difference 1 condition $P > 0.05$). With size differences 2 and 3, the accuracy was at chance level. The average latencies tended to increase with augmenting

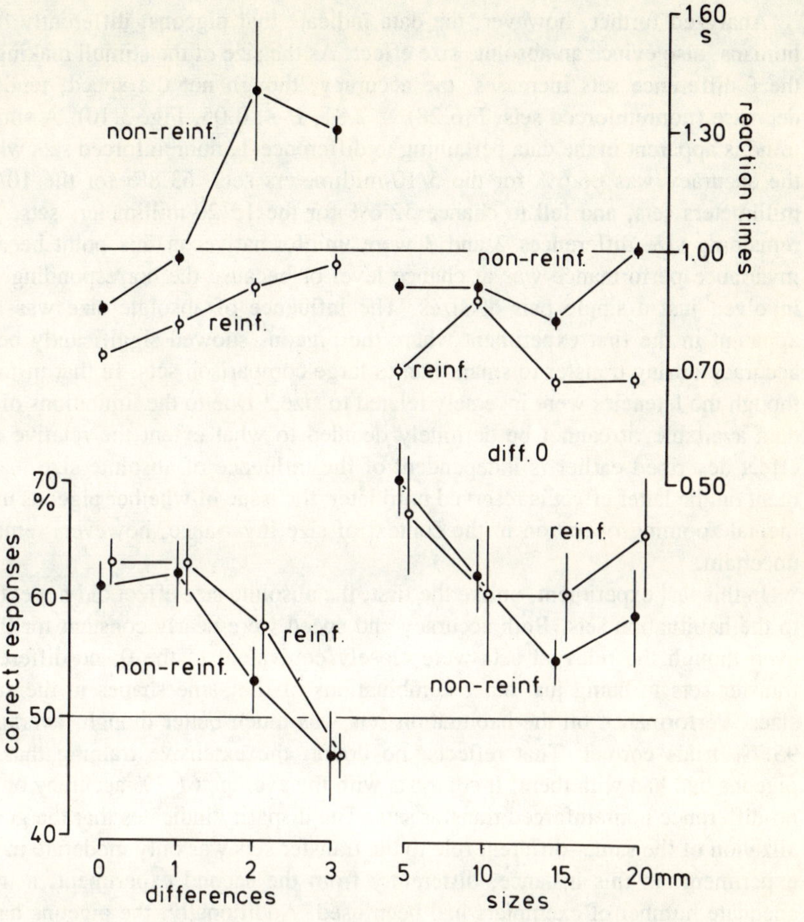

FIG. 3.10. Average accuracy and average speed both with standard deviations on transfer sets as a function of size differences between sample and comparisons (left) and for nodifference sets only as a function of the actual size of the stimulus shapes (right). Function of the actual size of the stimulus shapes (right).

size differences. A multivariate analysis of variance (Dixon & Brown, 1983; the latency data was entered untransformed, the accuracy data was linearized according to a suggestion by Pachella, 1974) resulted in an $F(6,28) = 6.58$, $P < 0.01$, indicating that increasing relative size differences between the comparanda influences the pigeons' size-invariance performance. To this extent pigeons would appear to behave like the human subjects of Bundesen and Larsen (1975), the data seemingly supporting the intervention of a mental zooming operation.

Analyzed further, however, the data indicate that pigeons, differently from humans, also evince an absolute size effect. As the size of the stimuli making up the 0-difference sets increases, the accuracy, though not the speed, tends to decrease (nonreinforced sets, $F(6,28) = 2.81$, $P < 0.05$, Fig. 3.10). A similar trend is apparent in the data pertaining to difference-1, nonreinforced sets where the accuracy was 66.6% for the 5/10-millimeters sets, 63.8% for the 10/15-millimeters sets, and fell to chance 52.6% for the 15/20-millimeters sets. The remaining size differences 2 and 3 were uninformative on this point because invariance performance was at chance level or because the corresponding sets involved just a single pair of sizes. The influence of absolute size was also apparent in the first experiment where the pigeons showed significantly better accuracy during transfer to small than to large comparison sets. In that instance though the latencies were inversely related to size.[1] Due to the limitations of the data available, it cannot be definitely decided to what extent the relative size effect described earlier is independent of the influence of absolute size. Comment on the latter effect is reserved until later; the issue of whether pigeons use a mental zooming operation in the context of size invariance, however, remains uncertain.

In this last experiment, unlike the first, the absolute size effect did not extend to the habituation sets. Both accuracy and speed were nearly constant for them even though the relevant sets were closely equivalent to the 0, no-difference transfer sets in being just other combinations of the same shapes in the same sizes. Performance on the habituation sets was much better though, averaging 95.7% trials correct. That reflects, no doubt, the extensive training that the pigeons had had with them. It contrasts with the average 61.2% accuracy on the no-difference nonreinforced transfer sets. The disparity indicates that the generalization of the same–different rule to the transfer sets was only moderate in this experiment. In this instance, differently from the second experiment, a more adequate number of exemplars had been used. Additionally, the pigeons had it easier because they knew all the shapes in all sizes that were used in the transfer test. It is suspect that the multifariousness of the size-invariance task (the sample as well as comparison shapes occurred in 4 sizes) reduced transfer to a comparatively moderate level.

---

[1]It is our impression from this and other experiments that the relation between response speed (reaction times) and response accuracy (correct choices) in discrimination experiments may be different in pigeons and humans. Whereas in the latter the two variables are, as a rule, inversely related (speed-accuracy tradeoff, Pachella, 1974), in pigeons accuracy seems to be an inverse-U function of speed. A given task may predominantly sample the falling leg or the rising leg or actually the apex of the function. The response latency for optimal accuracy also seems to vary with tasks and individuals. However, our pigeon experiments have always been of the self-paced variety, whereas human experiments usually involve constraints on both speed and accuracy imparted through instructions.

DISCUSSION

The results of the first and, less drastically, the third experiment, demonstrate that without ad hoc training pigeons are at least partially capable of size invariance in pattern-recognitions tasks; that is, they spontaneously recognize the equivalence or nonequivalence of shapes regardless of size differences between them, without having been specifically taught this skill. They can do so when sorting novel shapes discrepant in size according to a relational same–different rule. This is not at all to say that pigeons habitually disregard size when judging the equivalence of shapes, but only that they can do so, when forced, to at least some extent. Having to solve the oddity-from-sample problem when faced with comparison stimuli, both of which differed in scale from the samples, they clearly had no choice except to ignore the size differences. In fact, our results provide much suggestive evidence that the pigeons oddity performance was disturbed by having to do so. However that might be, the size-invariance performance of pigeons in our tests must be classed as comparatively modest. Pigeons may, after all, not be particularly predisposed to show size invariance in situations where size differences of objects are not correlated with distance differences. As pointed out at the beginning of this chapter, there may be restricted demand in the pigeon's environment for a size-compensating mechanism that is independent of viewing distance variations.

It would be risky in our opinion to try to predict from our results whether pigeons will show size invariance or not in pattern discrimination or recognition tasks of different design. There is much experience in the human perception field that to elicit one or the other invariance performance from subjects it is necessary to impart quite precise instructions. The reason is that human subjects can be simultaneously aware, for example, that two stimuli are the same shape but different size. Whether their behavior is guided by either the first or the second circumstance, or actually by neither or both, depends on the precise context both past and present. Invariance-governed responding is an option; it is not an obligation. Pigeons, at least a priori, cannot be expected to be differently organized. Unfortunately it is more difficult to impart precise instructions to animal subjects than to human subjects, misunderstandings being of common occurrence with the former. Negative evidence concerning this or that behavioral capacity of animals must be necessarily viewed with skepticism until the evidence is overwhelming. It should be obvious that animals might behave according to different decision rules in situations where options are less constrained and attached to other payoffs than in the present experiments.

Relating to size Cerella (this volume) reports results obtained with a successive conditioning technique that, at least superficially, are in agreement with ours. His pigeons generalized their responding to shapes differing in size by factors larger than 0.6 and smaller than 1.4 from the original positive training shapes. Our pigeons coped with size differences involving factors down to 0.5

and up to 2. Note though that the procedure Cerella used contains no element analogous to our habituation trials that would have informed the pigeons that size variations were irrelevant for the task at hand. Having the negative training shapes systematically vary in size could have been used to devalue size as a cue. During test trials the options of Cerella's pigeons were wide open; there was no forced choice. Pigeons have been shown to be capable of discriminating quite small size differences (Hodos, Bessette, Macko, & Weiss, 1985; Schwabl & Delius, 1984). A pigeon might therefore be aware of the identity of the stimulus shapes but recognize the aberration of the stimulus sizes. Nonresponding might easily be due to the latter factor, but it does not negate the former. This comment of course applies even more to the negative evidence that is reported relating to other invariances. The shape of generalization gradients (and that is what Cerella assesses) is known to be very sensitive to procedural variations (see, for example, Honig & Urcuioli, 1981; Jenkins, Pascal, & Walker, 1958). Excluding the fact that, for procedural reasons, he could not test spontaneous invariance with respect to novel shapes, there is however no appreciable dissension between Cerella's results and ours concerning size invariance. Blough (this volume), using a different paradigm, shows that pigeons judge as more similar large and small rectangles, and again large and small triangles than large triangles and large rectangles or small triangles and small rectangles. His data incidentally also provide evidence that pigeons can additionally command other invariances besides size invariance.

Leaving aside for a moment the size-invariance issue, some results of our first and third experiment suggested that, regardless of whether pigeons have to deal with size disparities or not, larger sized stimuli tend to yield a poorer oddity-from-sample performance. The experiments do not illuminate why that should be. Because this difficulty of pigeons has also been noted, although not absolutely consistently present, in some other unpublished experiments of ours, it might be worth considering a few possibilities. Just before pecking, pigeons view the target binocularly from a standard distance (6 to 10 centimeters; Hodos, Bessette, Macko, & Weiss, 1985).[2] Due to the lateral position of the eyes, their binocular visual field is restricted (Jahnke, 1984). It could be that they have difficulties in encompassing larger stimuli within that field. Estimates based on the known aperture of the field and of the distance at which pigeons fixate targets before pecking indicate, though, that 20-millimeters diameter stimuli should be well within binocular view. The areas of retina on which the binocular sector images, the so-called red field, is, however, not homogeneous (Emmerton,

[2]It is important to stress that while performing in the size-invariance tasks our pigeons did not compensate stimulus-size differences with viewing-distance adjustments that would have standardized the retinal images. In the first experiment fixation distances varying over a factor of 4 would have been necessary. Observations showed instead that their prechoice viewing distances fell within those given by Hodos, Bessette, Macko, and Weiss (1985), that is 8 ± 2 cm from eye to key, independently of stimulus size.

1983a). There might be a smaller area of optimal vision that could be exceeded by the image of the larger stimuli. In the context of pecking, pigeons normally need to discriminate acutely objects they can potentially swallow. The hypothetical functional frontal fovea might thus have quite a reduced aperture. Connected with that is the fact that, because they attempt to grasp them, pigeons relate stimuli projected on keys to food (LaMont & Zeigler, 1984). They should have an aversion to pecking at stimuli that are too large to swallow. That might interfere with discriminative responding. Some recent observations by R. Jäger (personal communication) indicate that pigeons tend to peck at the edges of large stimuli, which probably reflects the natural behavior of attempting to break off manageable pieces from large edible chunks. Jenkins and Sainsbury (1970) noted earlier that pigeons had a tendency to selectively peck at small features of rewarded stimuli. Jäger reports that, with 20-millimeter diameter stimuli within a 25-millimeter diameter key, pigeons often attempt to grasp the background annulus. All this might lead them to attend to features that are irrelevant to the discriminative problems at hand when dealing with large stimuli. A further possibility, completely unrelated to the foregoing, is that the greater luminous flux associated with the large stimuli might have dazzled the pigeons. In another experiment higher light intensities were identified as interfering with accurate pattern recognition (Lombardi & Delius, in preparation, see also Emmerton, 1983b). Small stimuli, incidentally, can create difficulties because of image degradation due to the optics of the projection system and eventually because they surpass the resolution of the pigeon eye.

Again with regard to size invariance, it is necessary to consider the suggestion that it is simply an instance of stimulus generalization. If the suggestion is meant as a descriptive label, as an effort to place the phenomenon within a set of better known behavioral effects, then there can only be some mild objections. One such objection is that invariance experiments typically examine the maintenance of a discrimination between, and not just the maintained responding to, stimuli on one stimulus dimension (shape, in this case), whereas the stimuli are varied in opposite directions on another dimension (size, in this case). Because it can be argued, however, that discrimination is nothing but differential responding, the objection is probably weak. But it does point to the fact that the particular kind of generalization involved in invariance experiments is, as far as we are aware, largely unexplored. Experiments on the effects of interdimensional and extradimensional training on generalization gradients (see Mackintosh, 1977) that superficially may seem relevant turn out on closer examination not really to address or even allow easy extrapolation to the issues raised by the invariance paradigm. The only really relevant experiment on pigeons, besides our own, interestingly makes no reference to stimulus generalization (Wright, 1976; the experiment concerns wave-length discriminations as a function of light-intensity variations).

The second mild objection is that, in the strictest version of the invariance paradigm as employed by us in the first experiment, the subjects were forced to apply a relational-discrimination rule to shapes to which they had not been conditioned before. Conventionally, stimulus generalization gradients only develop through conditioning to particular stimuli (see Rilling, 1977). So, what gradients could our subjects have resorted to when solving the task with novel shapes? Again, the objection is weak because contemporary theory (Blough, 1969) allows that any stimulus, even if it has not been involved in any conditioning, maps on an internal graded representation. This neurally coded gradient, however, is quite hypothetical. The theory entails further constructs, namely decision criteria, before it can predict behavior. Decision criteria are thought to arise through conditioning. In the invariance paradigm these would also need to generalize along the invariance dimension. As it stands, the theory does not envisage such a step. But all this means going beyond the simple issue of descriptive ascription of invariance to stimulus generalization and entering the complex realm of speculation of how generalization processes might intervene in invariance performance.

Even so, a provisional, purely formal amalgamation of the invariance with the generalization phenomena should perhaps not be resisted too strenuously. It may well be a heuristically productive way to proceed. The framework offered by the present experiments (and also Cerella's, this volume) suggests several obvious experiments along the generalization-gradient line of thought. For instance, it may be revealing to measure the gradients across the discrimination dimension (shape, in the present instance) as probe stimuli are displaced on the invariance dimension (size, in the present case). Could the invariance performance obtained in an equivalent situation be derived without strain from the gradients so measured? But such experiments need to be done if the assimilation of phenomena is to be based on facts rather than on wishes.

If, however, the attribution of invariance to generalization is meant to be an explanation, objections are strong. It will not do to explain something with something that itself needs explaining. The fact that the concept of stimulus generalization is familiar to the practitioners of animal-learning research should not be allowed to obscure the fact that the theories supposed to account for the very basic stimulus-generalization phenomena are not at all deeply developed. Apart from the fact that at present not one of them seems to be capable of explaining even a modest range of facts (Blough, 1983), none makes an effort to specify the information-processing operations underlying the generalization process such that, say, an artificial-vision designer could benefit from them when having to equip his system with size-invariance capabilities, or for that matter, with stimulus-generalization behavior. It can be no chance that those most closely concerned with invariance phenomena, the students of human perception, the neurophysiologists of sensory systems, and the engineers of artificial sensing systems, do not even care to make a passing reference to generalization gradients in their writings. Rather, the hunt is on for the more concrete information-processing operations that are capable of invariance performance or actually yield invariance. Analogical/propositional representations, transform algorithms, sequential/parallel processors, neuronal filtering are the constructs that are discussed (see, for example, Cooper, 1982; Glünder, 1986; Reitböck & Altmann, 1984; Perret, Smith, Potter, Mistlin, Head, Milner, & Jeeves, 1984). At that level of analysis, the stimulus-generalization framework as it exists at present is just far too global, too vague, too descriptive to be of any real use.

ACKNOWLEDGMENTS

The work was supported by the Deutsche Forschungsgemeinschaft through its Sonderforschungsbereich 114 from which C. M. Lombardi also received a visitor's grant. Her present address is Departamento de Biologia, Facultad de Ciencias Exactas y Naturales, Universidad de Buenos Aires, R A 1428, Buenos Aires, Argentina. J. D. Delius' present address is: Allgemeine Psychologie, FG Psychologie, Universität, D 7750 Konstanz, F.R. Germany. We warmly thank G. Keim, A. Lohmann, and D. Hagenkötter for absolutely essential and very patient assistance and Dr. J. Emmerton for many useful suggestions and editing. Comments by Prof. J. E. R. Staddon and Prof. R. J. Herrnstein helped to improve several passages. Dr. M. L. Commons did much to improve the English.

REFERENCES

Besner, D. (1983). Visual pattern recognition: Size preprocessing re-examined. *Quarterly Journal of Experimental Psychology, 35,* 209–216.

Besner, D., & Coltheart, M. (1976). Mental size scaling examined. *Memory and Cognition, 4,* 525–531.

Blough, D. S. (1969). Generalization gradient shape and summation in steady state test. *Journal of the Experimental Analysis of Behavior, 12,* 91–104.

Blough, D. S. (1983). Alternative accounts of dimensional stimulus control. In M. L. Commons, R. J. Herrnstein, & A. R. Wagner (Eds.), *Quantitative analysis of behavior: Vol. IV, Discrimination processes* (pp. 59–72). Cambridge, MA: Ballinger.

Bundesen, C., & Larsen, A. (1975). Visual transformation of size. *Journal of Experimental Psychology: Human Perception and Performance, 1,* 214–220.

Bundesen, C., Larsen, A., & Farrel, J. E. (1981). Mental transformation of size and orientation. In J. Long & A. Baddeley (Eds.), *Attention and performance* (Vol. IX, pp. 279–294). Hillsdale, NJ: Lawrence Erlbaum Associates.

Cerella, J. (1979). Visual classes and natural categories in the pigeon. *Journal of Experimental Psychology: Human Perception and Performance, 5,* 68–77.

Cerella, J. (1982). Mechanisms of concept formation in the pigeon. In D. J. Ingle, M. A. Goodale, & R. J. W. Mansfield (Eds.), *Analysis of visual behavior* (pp. 241–260). Cambridge, MA: MIT Press.

Cooper, L. A. (1982). Internal representations. In D. R. Griffin (Ed.), *Animal mind, human mind* (pp. 145–158). Berlin: Springer.

Corballis, M. C., & Beale, I. L. (1976). *The psychology of left and right.* Hillsdale, NJ: Lawrence Erlbaum Associates.

Delius, J. D. (1985). Cognitive processes in pigeons. In G. d'Ydewalle (Ed.), *Cognition, information processing and motivation* (pp. 3–18). Amsterdam: Elsevier.

Delius, J. D. (1986). Complex visual information processing in the pigeon. In V. D. Iljitschew & V. M. Gavrilov (Eds.), *Acta of the XVIII congressus internationalis ornithologici* (pp. 804–811). Nauka, Moscow.

Delius, J. D., & Hollard, V. D. (1987). Rotational invariance of shape recognition in forebrain lesioned pigeons. *Behavioural Brain Research, 23,* 251–259.

Delius, J. D., & Nowak, B. (1982). Visual symmetry recognition by pigeons. *Psychological Research, 44,* 199–212.

Dixon, W. J., & Brown, M. B. (1983). *Biomedical computer programs.* Berkeley: University of California Press.

Emmerton, J. (1983a). Functional morphology of the visual system. In M. Abs (Ed.), *Physiology and behaviour of the pigeon* (pp. 221–243). London: Academic Press.

Emmerton, J. (1983b). Vision. In M. Abs (Ed.), *Physiology and behaviour of the pigeon* (pp. 245–266). London: Academic Press.

Fellows, B. J. (1967). Chance stimulus sequences for discrimination tasks. *Psychological Bulletin, 67,* 87–92.

Glünder, H. (1986). Neural computation of inner geometric pattern relations. *Biological Cybernetics, 55,* 239–251.

Herrnstein, R. J. (1985). Riddles of natural categorization. *Philosophical Transactions of the Royal Society, London B, 308,* 129–144.

Herrnstein, R. J., Loveland, D. H., & Cable, C. (1976). Natural concept in the pigeon. *Journal of Experimental Psychology, Animal Behaviour Processes, 2,* 258–302.

Hodos, W., Bessette, B. B. Macko, K. A., & Weiss, S. R. B. (1985). Normative data for pigeon vision. *Vision Research, 25,* 1525–1527.

Hollard, V. D., & Delius, J. D. (1982). Rotational invariance in visual pattern recognition by pigeons and humans. *Science, 218,* 804–806.

Honig, W. K., & Urcuioli, P. J. (1981). The legacy of Guttman and Kalish (1956): 25 years of research on stimulus generalization. *Journal of the Experimental Analysis of Behavior, 36,* 405–445.

Hord, R. M. (1982). *Digital image processing of remotely sensed data.* New York: Academic Press.

Howard, J. H., & Kerst, S. M. (1978). Directional effects of size changes on the comparison of visual shapes. *American Journal of Psychology, 91,* 491–499.

Jahnke, H. J. (1984). Binocular visual field differences among various breeds of pigeons. *Bird Behaviour, 5,* 96–102.

Jenkins, H. M., & Sainsbury, R. S. (1970). Discriminative learning with the distinctive feature on positive or negative trials. In D. J. Mostofsky (Ed.), *Attention: Contemporary theory and analysis* (pp. 239–273). New York: Appleton-Century-Croft.

Jenkins, W. O., Pascal, G. R., & Walker, Jr., R. W. (1958). Deprivation and size generalization. *Journal of Experimental Psychology, 56,* 274–277.

Kubovy, M., & Podgorny, P. (1981). Does pattern matching require the normalization of size and orientation? *Perception and Psychophysics, 30,* 24–28.

LaMont, B. C., & Zeigler, H. P. (1984). Grasping in the pigeon (Columba livia): Stimulus control during conditioned and consummatory responses. *Animal Learning and Behavior, 12,* 223–231.

Larsen, A. (1985). Pattern matching: Effects of size ratio, angular difference in orientation, and familiarity. *Perception and Psychophysics, 38,* 63–68.

Lea, S. E. G. (1984). Complex general process learning in nonmammalian vertebrates. In P. Marler & H. S. Terrace (Eds.), *The biology of learning* (pp. 373–397). Berlin: Springer.

Lea, S. E. G., & Ryan, C. M. E. (1983). Feature analysis of pigeons acquisition of concept discrimination. In M. L. Commons, R. J. Herrnstein, & A. R. Wagner (Eds.), *Quantitative analyses of behavior: Vol. IV, Discrimination processes* (pp. 239–253). Cambridge MA: Ballinger.

Lohmann, A., Delius, J. D., Hollard, V. D., & Friesel, M. (in press). Discrimination of shape reflections and shape orientation by pigeons. *Journal of Comparative Physiology*.

Lombardi, C. M. (in press). Shape oddity recognition by pigeons is independant of shape orientation. *Revista Mexicana del Analysis de la Conducta*.

Lombardi, C. M., & Delius, J. D. (in preparation). *Pattern recognition invariance in pigeons: Outline, contrast and colour*.

Lombardi, C. M., Delius, J. D., & Hollard, V. D. (1986). Conceptualizacion de una relacion logica en palomas. *Revista Latinoamericana de Psicologia, 18*, 277–297.

Lombardi, C. M., Fachinelli, C. C., & Delius, J. D. (1984). Oddity of visual patterns conceptualized by pigeons. *Animal Learning and Behavior, 12*, 2–6.

Mackintosh, N. J. (1977). Stimulus control: Attentional factors. In W. K. Honig & J. E. R. Staddon (Eds.), *Handbook of operant behavior* (pp. 481–513). Englewood Cliffs, NJ: Prentice Hall.

Pachella, R. G. (1974). The interpretation of reaction time in information-processing research. In B. H. Kantowitz (Ed.), *Human information processing: Tutorials in performance and cognition* (pp. 41–82). Hillsdale, NJ: Lawrence Erlbaum Associates.

Perrett, D. I., Smith, P. A. J., Potter, D. D., Mistlin, A. J., Head, A. S., Milner, A. D., & Jeeves, M. A. (1984). Neurones responsive to faces in the temporal cortex: Studies of functional organization, sensitivity to identity and relation to perception. *Human Neurobiology, 3*, 197–208.

Pisacreta, R., Potter, C., & Lefave, P. (1984). Matching of varying size form stimuli in the pigeon. *Bulletin of the Psychonomic Society, 22*, 591–593.

Reitböck, H. J., & Altmann, J. (1984). A model for size- and rotation-invariant pattern processing in the visual system. *Biological Cybernetics, 5*, 113–121.

Rilling, M. (1977). Stimulus control and inhibitory processes. In W. K. Honig & J. E. R. Staddon (Eds.), *Handbook of operant behavior* (pp. 432–480). Englewood Cliffs, NJ: Prentice-Hall.

Schwabl, U., & Delius, J. D. (1984). Visual length discrimination in the pigeon. *Bird Behavior, 5*, 118–121.

Sekuler, R., & Nash, D. (1972). Speed of size scaling in human vision. *Psychonomic Science, 27*, 93–94.

Siegel, R. K., & Honig, W. K. (1970). Pigeon concept formation: Successive and simultaneous acquisition. *Journal of the Experimental Analysis of Behavior, 13*, 385–390.

Todrin, D. C., & Blough, D. S. (1983). The discrimination of mirror-image forms by pigeons. *Perception and Psychophysics, 34*, 397–402.

Towe, A. L. (1954). A study of figural equivalence in the pigeon. *Journal of Comparative and Physiological Psychology, 47*, 283–287.

Vaughan, W., & Greene, S. L. (1984). Pigeon visual memory capacity. *Journal of Experimental Psychology: Animal Behaviour Processes, 10*, 256–271.

Wildemann, D. G., & Holland, J. G. (1973). The effect of the blackout method on acquisition and generalization. *Journal of the Experimental Analysis of Behaviour, 19*, 73–80.

Wright, A. A. (1976). Bezold-Brücke hue shift functions for the pigeon. *Vision Research, 16*, 765–774.

4 Pigeons' Perception of Complex Movement Patterns

Jacky Emmerton
Ruhr-Universität Bochum, Psychologisches Institut, Federal Republic of Germany

Most of the research on pattern recognition in animals has concentrated on the role of static dimensions or spatial features in visual perception. The types of stimuli employed have typically been simple colored forms, more complex geometric shapes, or naturalistic photographic scenes. The undoubted importance of various static aspects of visual arrays in both human and animal perception is amply exemplified in the research of many of the other contributors to this volume. In the natural environment, however, many objects, particularly living ones, can be recognized by their movement characteristics.

The relative movement of an animal against a background that itself may not be totally motionless (e.g., swaying grass, sailing clouds) enhances the figure–ground distinction. This can enable a predator to detect its prey more easily, or conversely, prey, its predator. Another function of movement perception is to aid species identification. Species-typical movement patterns become particularly important when form and coloration of two groups of animals make them difficult to distinguish, or when information about an organism's shape is reduced to a minimum. The latter situation occurs when the animal is only visible at a distance as a small, dark speck. Examples of this use of motion patterns for species recognition can be found in any bird-watcher's handbook (Heinzel, Fitter, & Parslow, 1972; Robbins, Bruun, & Zim, 1966). Where the size, shape, and plumage coloration of related birds are similar, the distinctive flight patterns of different species provide an additional means of identification. For conspecifics, the pattern of movement performed by an individual can also inform others about its behavioral state, for instance, whether its display indicates that it is prepared to mate (Lorenz, 1958) or is aggressive and ready to attack.

The saliency of "biological motion" has also been demonstrated in the laboratory by Johansson (1973). Biological motion refers to the complex of coordinated rhythmic movements performed by different parts of the bodies of living organisms. Human observers were tested for their ability to recognize the movement patterns of other humans. The shape of a human figure was obscured and only a small number of light-reflecting spots, affixed to the joints, were visible to the observers. While immobile, these collections of dots remained meaningless, but as soon as the figure moved, it was easily interpreted as a human being walking or performing some other specifiable activity. Similarly, the pattern of movement could be used to distinguish gender (Kozlowski & Cutting, 1977) or to recognize a particular individual (Cutting & Kozlowski, 1977). Even very young infants showed a marked preference for looking at biological motion in contrast to mechanical oscillation of the same number of component light points (Fox & McDaniel, 1982).

From these examples from both ethology and psychology, one can surmise that the recognition of movement patterns should not only be salient to humans but could also have survival value to a visually guided animal such as the pigeon.

Single-cell recordings from the pigeon's visual system reveal that the majority of visually responsive units is optimally activated by moving stimuli (see Emmerton, 1983). Behavioral experiments show that these birds can learn to distinguish stationary from moving stimuli, irrespective of whether motion is real or apparent (as in the phi-phenomenon; Siegel, 1970, 1971). The pigeon's minimal velocity sensitivity (Hodos, Smith, & Bonbright, 1976) and differential sensitivity to speed of movement (Mulvanny, 1978) have also been measured psychophysically. Surprisingly, the movement thresholds obtained from pigeons are much poorer than those determined when humans are tested in comparable situations. In the preceding psychophysical experiments, the birds probably viewed the stimuli frontally because these stimuli were visible through the response keys. The pigeon, however, is relatively more sensitive to stimulus motion in its lateral than in its frontal visual field (Martinoya, Rivaud, & Bloch, 1983). The lateral field is generally associated with monocular viewing of more distant stimuli. Vision in the frontal field is binocular and is used for viewing nearby objects, for instance, seeds or response keys that the bird pecks. Experiments to date have therefore established that pigeons can detect movement per se. But there have been no studies of their ability to discriminate more complex patterns of movement.

To investigate the perception of movement patterns in these animals, stimuli were used that showed somewhat complex motion but that were nevertheless simple to generate. This also had the advantage that their input parameters could be easily specified.

DISCRIMINATION OF MOVING LISSAJOUS FIGURES

Pigeons were trained in a "go/no go" procedure to discriminate moving Lissajous figures that were successively presented on a single, 2.5 centimeter-diameter response key in a Skinner box. Lissajous figures were generated by applying two sine waves to the X and Y channels of an oscilloscope and projecting this display via a camera lens onto the key. The stimuli were thus reduced in size to project within a 7-millimeter^2 area in the center of the key. The particular type of movement pattern produced depends on the input frequencies and their ratio. When the frequency ratio is fractionally offset from integer values, the phase relation of the two sine waves, and thus the stimulus configuration, is continuously changing. This generates patterns that vary in a cyclically repetitive fashion.

Each experimental session consisted of 40 trials. The standard presentation time for positive ($S+$) and negative ($S-$) stimuli was 20 seconds each, with order of presentation being quasirandom. Pecks to the $S+$ were rewarded on a gradually increasing variable–interval (VI) schedule (maximally VI 10 seconds), whereas each peck to the $S-$ initiated an extension of stimulus presentation. In the early stages of training this extension period was increased to 10 seconds. Thus, if the bird continued to peck the $S-$, this stimulus could stay on indefinitely. Only pecks during the initial 20 seconds of both $S+$ and $S-$ trials were scored to compute discrimination accuracy, the percentage of pecks to the correct stimulus.

The initial experiment was designed to mimic the kind of situation with which an ornithologist is confronted when he or she identifies a bird's species on the basis of seeing a small distant spot in flight. Here the information normally available about a bird's complex shape is minimal. Similarly, all that was visible in the experimental display that the pigeons saw was a single small dot that described one of two repetitively changing trajectories. The movement patterns of the dot were generated with low input frequencies. When the frequency ratio was slightly offset from 1 to 1 (with actual frequencies of 2.8 to 2.7 Hz), the dot moved along a diagonal line. The pathway then became elliptical, circular, elliptical, and diagonal again before reversing through this changing trajectory (Fig. 4.1). It actually took 12 seconds to complete these cyclical changes. This stimulus had to be discriminated from another in which the dot's trajectory gradually varied between a U, a figure 8, and an inverted U. For this stimulus the frequency ratio was approximately 1 to 2, given by input frequencies of 1.3 to 2.7 Hz, and the cycle duration was 4.6 seconds. Thus, for both stimuli, the various trajectories were seen repeatedly within a single trial. For 3 out of 6 birds the 1 to 1 ratio stimulus was designated as $S+$ and the 1 to 2 ratio stimulus as $S-$. For the other 3 animals these contingencies were reversed.

Five of the birds trained on this initial task successfully learned to discriminate the two movement patterns at a level of more than 90% correct responses

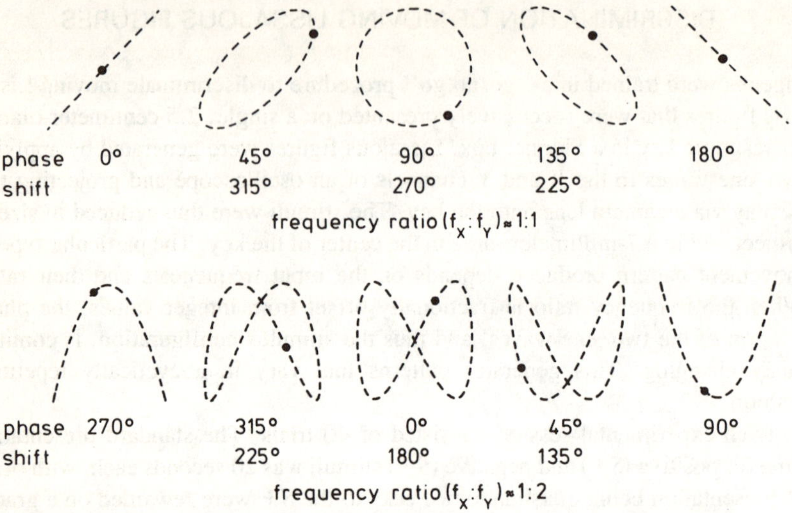

FIG. 4.1. The stimulus in the first experiment was a single moving dot. *S+* and *S−* differed in the dot's cyclically variable trajectories. The ratios of low sinusoidal frequencies, f_X and f_Y, were slightly offset from integer values, which produced continuously changing movement patterns by means of gradual phase shifts. The dot's trajectories are shown for momentary phase differences to give an impression of the stimulus movement patterns.

(Fig. 4.2). Catch trials (4 positive and 4 negative) were interspersed in the final training sessions. In these trials animals were tested under extinction to ensure that they were responding under stimulus control and not just reacting to the reinforcement contingencies. The one animal that failed to discriminate (about 60% correct responses), in spite of a fading technique being employed, pecked vigorously at all moving stimuli. Although this bird's performance improved slightly in subsequent experiments, it continued to show a lower and more erratic level of discrimination than the other birds. It was therefore excluded from the data analysis.

Although the visible stimulus lacked complex shape, the birds nevertheless mastered the discrimination of movement patterns quite quickly. They reached a discrimination criterion of more than 90% correct in roughly the same number of sessions as other pigeons, trained with a similar procedure, had needed to discriminate symmetric and asymmetric geometric shapes (Delius & Nowak, 1982). On each *S+* trial the initial peck occurred on average within 1 second of stimulus onset. A stimulus was therefore correctly identified when a given bird had seen only part of the dot's complex trajectory. Additionally, the stimulus phase at the start of each trial varied stochastically because trial initiation was timed independently from the generation of movement patterns. The birds generally responded steadily on $S+$ trials, with pecking being interrupted only by occasional reward periods. On the rare occasions when the animals responded to $S-$, latencies to first pecks were longer, averaging 1.6 seconds, and further responding was sporadic. Because all that was visible to the pigeons on each trial was a single, small dot in motion, the birds' successful acquisition of the discrimination task suggested that they had learned to differentiate two movement gestalts, i.e., they discriminated on the basis of movement pattern alone. This entailed temporal integration of the continuous displacement of the dot. To check whether there were any obvious differences between the two stimuli in the dot's location, continuous traces were plotted for 20-second periods corresponding to standard trial durations (Fig. 4.3). In each case, the dot traversed almost the whole area within which a stimulus was presented. Pattern discrimination therefore could not have been based on some gross difference in the dot's location, such as one stimulus covering only a central projection area on the pecking key and another one a more peripheral region. The slight differences in the dot's cumulative locations that obviously do exist between the two movement patterns would nevertheless still have to be integrated over time before pigeons could make use of the dot's position as a discriminative cue.

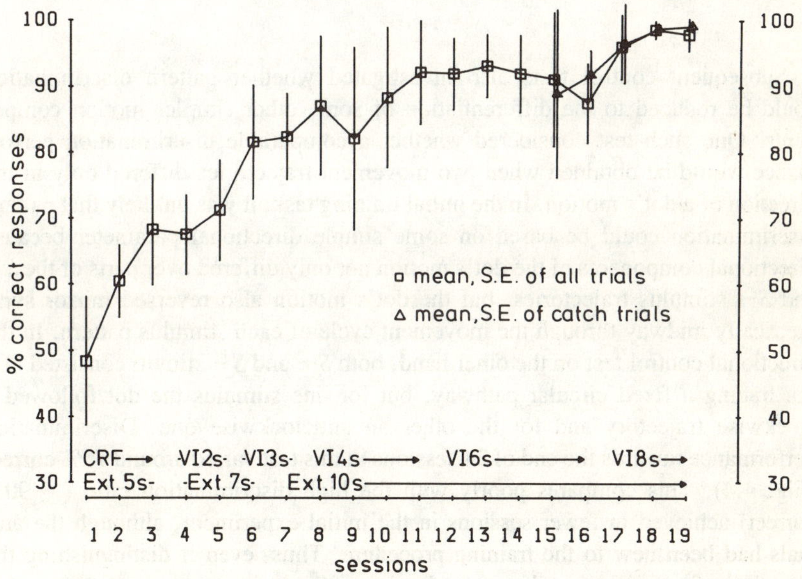

FIG. 4.2. Acquisition curve with mean percentage correct scores ± S.E. for five birds that learned to discriminate the moving-dot configurations. Results from interspersed catch trials are shown for the final sessions. Changes in the reinforcement schedule and duration of $S-$ extension are indicated on the abscissa.

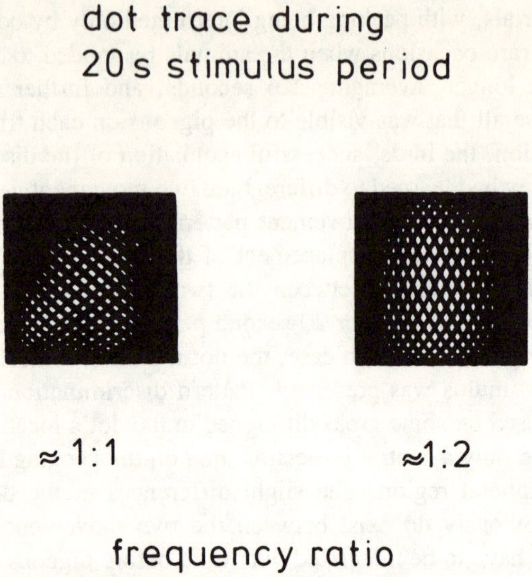

FIG. 4.3. Continuous traces of the trajectories of the moving-dot configurations used in the first experiment. The traces shown changing location of the dot during a 20 s stimulus period.

Subsequent control tests also investigated whether pattern discrimination could be reduced to the differentiation of some other simpler motion components. One such test considered whether a comparable discrimination performance would be obtained when two movement trajectories differed only in the direction of a dot's motion. In the initial training task, it was unlikely that pattern discrimination could be based on some simple directional parameter because directional components of the dot's motion not only differed over parts of the $S+$ and $S-$ stimulus trajectories, but the dot's motion also reversed mirror symmetrically midway through the movement cycle of each stimulus pattern. In the directional control test on the other hand, both $S+$ and $S-$ stimuli consisted of a dot tracing a fixed circular pathway, but for one stimulus the dot followed a clockwise trajectory and for the other an anticlockwise one. Discrimination performance towards the end of 24 sessions in this test varied around 75% correct (Fig. 4.4). This compares poorly with the high discrimination ratio (> 90% correct) achieved in fewer sessions in the initial experiment, although the animals had been new to the training procedure. Thus, even if distinguishing the direction of motion contributes to the perception of movement patterns, this component alone is insufficient to account in any simple way for the high level of accuracy with which motion patterns were discriminated. The suggestion that the pigeons distinguished motion gestalts rather than some single motion parameter was further supported by rotation and velocity control tests described later.

With real objects in the natural environment, both their shapes and movement patterns are often visible to the onlooker. This type of situation was modeled in a second experiment with another variety of Lissajous figures, consisting of contours in motion.

The same birds trained in the first experiment were transferred to a new discrimination of two different moving outline configurations. These movement patterns were generated with higher input frequencies. For one stimulus, the frequency ratio approximated 2 to 1 (184.6 to 92.7 Hz), whereas the other configuration was produced by a frequency ratio of nearly 3 to 1 (278.8 to 92.7 Hz). An impression of the continuous changes in stimulus shape is given by the selected momentary stages of a cycle depicted in Fig. 4.5. Cycle durations for these stimuli were 1.8 seconds and 1.9 seconds, respectively.

The five pigeons that had previously successfully discriminated the moving dot patterns also quickly reached an acquisition criterion of more than 90% correct responses in this moving-outline discrimination (Fig. 4.6). Geometrically, the stimuli are congruent with images, projected on a plane, of regularly shaped, rigid wire objects turning about a central horizontal axis. To human observers, the displays on the oscilloscope screen actually give an impression of depth and show spontaneous "reversals" in the direction of turning (Fisichelli, 1946; Philip & Fisichelli, 1945; see also Ullman, 1984). Whether the two-dimensional moving outlines are perceived as three-dimensional by pigeons cannot be ascertained from the present data. Nevertheless, because the stimuli showed continuous and repetitive variation in contour, a computational description of such stimuli is far from simple (cf. Hildreth, 1984). An efficient discrimination of the stimuli would seem to demand a temporal integration of the changes

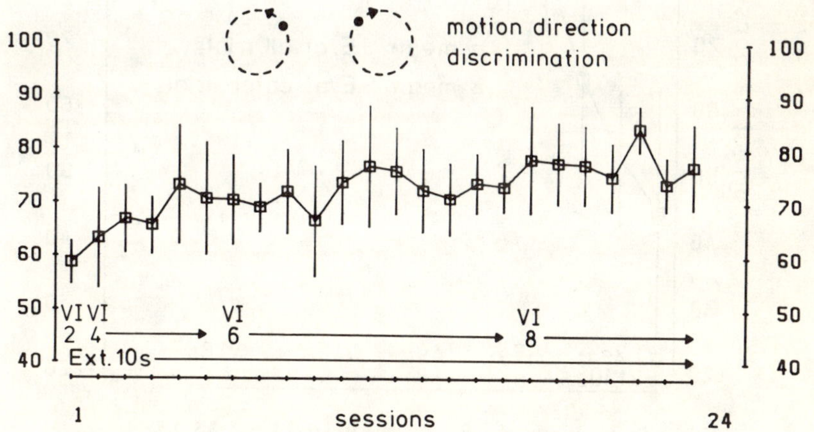

FIG. 4.4. Discrimination of the direction of a moving dot that moved along a constant circular trajectory in either a clockwise or an anticlockwise direction. Mean data ± S.E.s are shown.

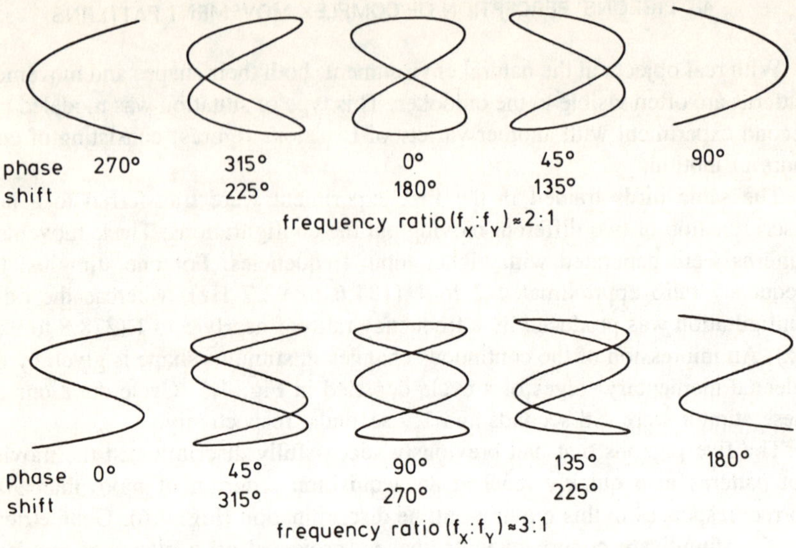

FIG. 4.5. Moving stimuli in the second experiment were closed outline figures. These were generated by high input frequencies.

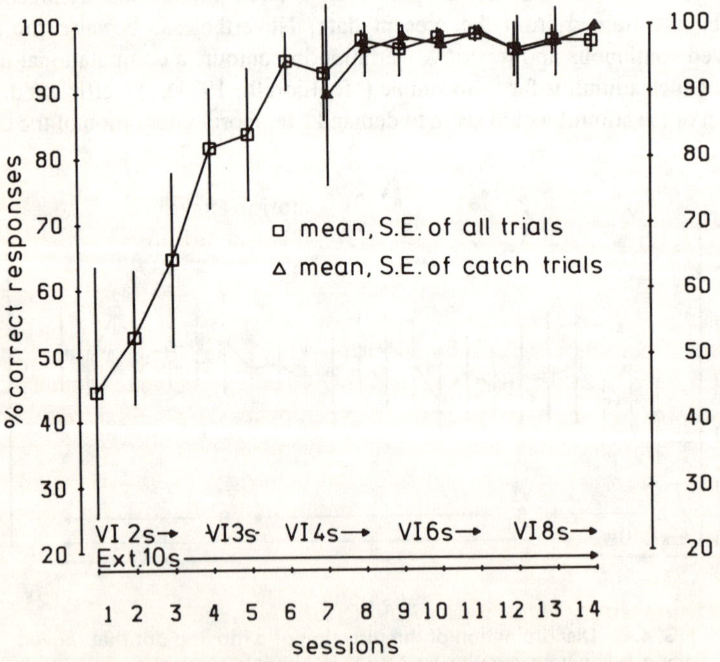

FIG. 4.6. Mean acquisition data for the discrimination of moving-outline stimuli.

in shape. Quick responding, after the correct stimulus is revealed by opening a shutter, would require that the pigeons recognize only part of the movement cycle as being a component of the whole configuration. On average, birds responded to the $S+$ within 1 second of stimulus onset and then pecked more or less continuously throughout the trial. Response latencies were quite consistent, independent of specific $S+$ configurations. The birds therefore did not seem to wait until any specific stage, or momentary contour, appeared before starting to peck. The latencies in this discrimination of moving outlines also concurred with those obtained in the previous discrimination of moving-dot trajectories.

To the human eye, the outline stimuli were easier to discriminate than the single moving-dot trajectories. The outline stimuli were richer in perceptual information about both shape and movement. Acquisition in the second experiment was also quicker than in the first. Because all birds were trained in the same order, however, this difference in learning rate might simply have been due to increased task familiarity. To test whether the difference in learning rates reflected perceptual factors, 10 new birds were tested in a counterbalanced design to discriminate both moving-dot and moving-outline patterns. They were first familiarized with the "go/no go" procedure for 9 sessions with static patches of red and white light. During this time the VI schedule was gradually lengthened to VI 10 seconds. Within 3 sessions the birds were discriminating with at least 90% accuracy. Half the animals were then initially trained to discriminate moving-dot patterns and the other half moving outlines. Stimulus conditions were reversed in the second half of the experiment. For this acquisition comparison, the discriminative stimuli were always the type that included elliptical stages (with a nearly 1 to 1 frequency ratio) or the type that included a figure-8 stage (with frequency ratios of approximately 1 to 2 or 2 to 1). The input frequencies used to generate outline stimuli were set higher than those producing dot stimuli. For all these stimuli, cycle durations were almost equal at 4.0 to 4.5 seconds.

In both halves of the experiment the discrimination of moving outlines was learned significantly more quickly than that of moving-dot trajectories ($F(1,8) = 22.5, p < 0.01$ and $F(1,8) = 8.8, p < 0.05$, respectively; Fig. 4.7). In humans, separate channels for the analysis of the shape and the movement of visual stimuli have been inferred from psychophysical experiments (Tolhurst, 1973). These two channels have been related to neurophysiologically distinct X- and Y-cells in the mammalian visual system. In the avian optic tectum, there are cells that are tuned to spatial- and temporal-frequency bands (Hardy & Jassik-Gerschenfeld, 1979). Presumably, a stimulus consisting of a moving contour more effectively activates neural units.

Rotational Invariance with Moving Stimuli

The human visual system can cope with remarkable ease with a number of aspects of perceptual invariance. Human observers easily recognize a pattern or object as remaining the same "thing" in spite of many transformations in the

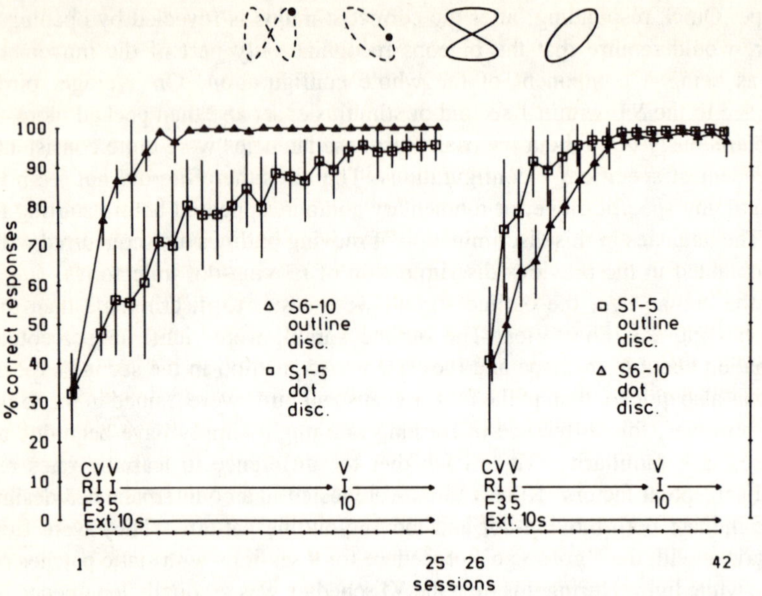

FIG. 4.7. Comparison of rates of learning to discriminate moving-dot or moving-outline stimuli for 10 subjects, S1–10. The types of continuously variable movement patterns used are represented by single, momentary configurations.

optic array impinging on the retina. These changes include translation, rotation, size, occlusion, and distortion. Regular motion in a two-dimensional array can produce an impression of three-dimensional structure for humans. As previously mentioned, this is often the case with outline Lissajous figures. Computational theories of pattern recognition derive an object's 3-D structure from motion invariants by applying an assumption of rigidity, or least structural change. Algorithms incorporating this operating principle have successfully modeled some of the phenomena of human motion perception (Ullman, 1984). Whether three dimensionality is perceived or not, motion invariance implies perceiving some unity in an optic array that undergoes regular temporal transformations. Results from the experiments described earlier suggest that pigeons are capable of at least this level of motion invariance.

With static shapes there is evidence that pigeons as well as humans can recognize the essential equivalence of patterns whose size (Cerella, this volume; Lombardi & Delius, this volume) or contrast (Lombardi & Delius, in preparation) has been changed. Furthermore, pigeons are capable of rotational invariance after alterations in stimulus orientation (Hollard & Delius, 1982; but see Cerella, this volume). With the same pigeons previously trained in the first two experiments, the effect of orientation changes on the recognition of movement patterns was investigated. A Lissajous figure can be axis rotated through 90° by simply switching X and Y inputs to the oscilloscope. This was done on test trials included within sessions. Immediately following discrimination training on a pair of outline stimuli, the orientations of which had remained constant throughout 32 noncatch and 8 catch trials, generalization to reoriented stimuli was tested in 3 further sessions. In these sessions, catch trials were replaced by rotation trials. On the remaining trials stimulus orientation was standard and responding was reinforced as before. The 4 kinds of trials from the 3 training and the 3 test sessions were accordingly designated *noncatch, catch, nonrotation,* and *rotation.* Results for these 4 types of trials were computed separately. The discrimination performance, in terms of the percentages of correct responses, did not significantly differ among all 4 conditions ($F(3,44) = 0.63, p > 0.05$) (Fig. 4.8). The absolute numbers of responses made per session on 4 trials of each type were also compared. The numbers of pecks to rotated $S+$ stimuli decreased steadily over the 3 sessions. The response rates to these stimuli differed significantly from those to nonrotated stimuli presented in the same sessions but on reinforced trials ($F(1,20) = 29.7, p < 0.001$). Response rates between noncatch and catch trials, in which stimulus orientations remained the same, did not differ ($F(1,20) = 0.6, p > 0.05$). Response rates were consistently low ($F(3,44) = 0.157, p > 0.05$) on all types of $S-$ trials.

The results with rotated moving-outline stimuli can be summarized as follows: The birds still discriminated $S+$ and $S-$ stimuli accurately on rotation trials. They must therefore have immediately recognized that rotated and nonrotated configurations were similar. At the same time the decreased responding to rotated $S+$ stimuli implies that the animals learned that the new stimulus orientation was associated with nonreward. This decrease in responding is typically obtained when stimulus generalization along a given dimension is tested under extinction over a number of sessions (Guttman & Kalish, 1956).

The same animals were subsequently retrained on the discrimination of the moving-dot configurations used in the first experiment. As with the outline stimuli, they were then tested in 3 further sessions that included trials with rotated stimuli. Exchanging the X and Y frequencies produced a 90° axis rotation in the figure-8-type stimulus ($\approx$ 1 to 2 frequency ratio). Exchanging the input frequencies of the ellipse-type stimulus ($\approx$ 1 to 1 ratio) led only to a change in the direction of spin of the moving dot, because this configuration already included stages in which the orientation of the dot's track differed by 90° (Fig. 4.1). Normally, the dot described an anticlockwise track when the figure changed between the 0 and 180° phases and clockwise in the other half of the cycle. These directions were reversed on "rotation" trials with this figure. For 3 birds this figure was the $S+$ and for 2 the $S-$.

Overall, discrimination accuracy on the rotation trials decreased with moving-dot stimuli (nonrotated vs. rotated: $F(1,20) = 28.2, p < 0.001$) (Fig. 4.8). This was due to changes in responding to the truly rotated figure-8 stimulus. This

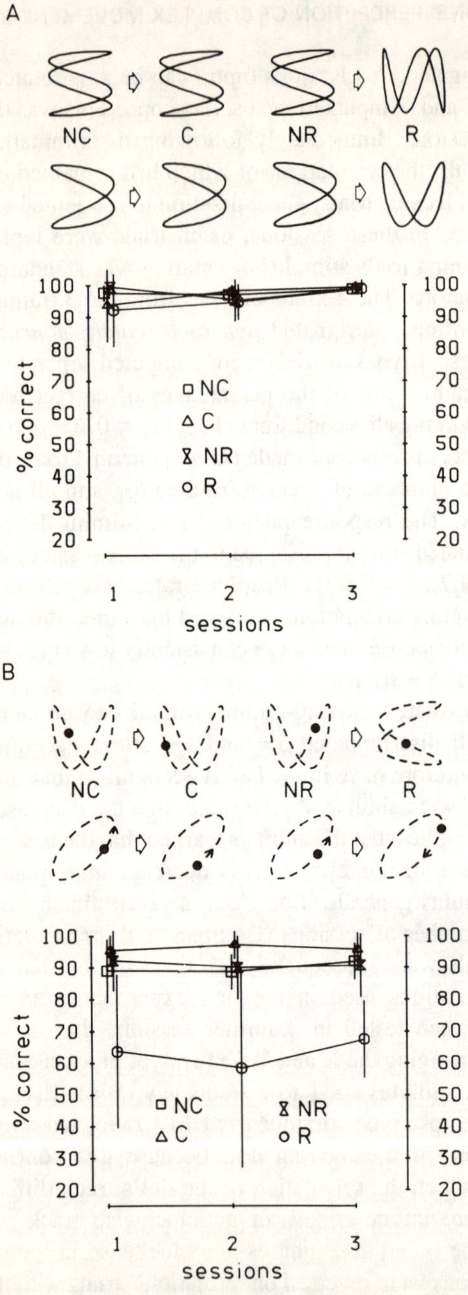

FIG. 4.8. Comparison of mean discrimination accuracy on tests for rotational invariance with (a) moving-outline and (b) moving-dot stimuli. Data from noncatch (NC), catch (C), nonrotation (NR), and rotation (R) trials were computed separately. Examples of stimuli for each trial type are given. In rotation trials, either axis orientation of the stimulus configuration or direction of spin of the moving dot was altered.

stimulus was the $S-$ for 3 animals. On rotation trials, these birds no longer withheld pecking to the $S-$ so that their response rates increased. This contrasted with the consistently low response rates of the same birds to rotated outline $S-$ stimuli ($F(1,10) = 53.07, p < 0.001$). On all $S+$ rotation trials with the figure-8-type dot stimulus, one bird pecked considerably less frequently than on nonrotation trials. The other bird, the only one to maintain a high discrimination score on this rotation test, showed little change in its response rate. The other stimulus including the elliptical trajectory had its direction of spin, rather than its axis orientation, altered on "rotation" trials. Response rates on these trials were similar to those on corresponding $S+$ or $S-$ catch trials.

Contrasting results were therefore obtained when moving-outline and moving-dot stimuli were rotated. The birds demonstrated rotational invariance with mobile outline patterns, immediately recognizing the rotated figures as being equivalent to those in the familiar nonrotated standard orientation used throughout training, and treating these stimuli as similar, in spite of orientational changes (cf. Hollard & Delius, 1982). The same birds, however, discriminated the moving-dot patterns poorly on rotational trials and thus failed to manifest rotational invariance in this test. A similar difficulty in recognizing mobile dot figures has been reported with human observers when Johansson's dot arrays of a walking person were displayed upside down (rotated 180°) on a video. Although they still recognized seeing some kind of biological motion, the movement appeared strange and they found it more difficult to specify what activity was portrayed (Fox & McDaniel, 1982; Sumi, 1984). So far no equivalent demonstration has been performed with human subjects using moving-outline pictures. But comparing the results with the different moving Lissajous figures implies that contour information, as provided in the outline stimuli, must be available for rotational invariance to be spontaneously manifested. Of course, the static geometric shapes that have been used to test rotational invariance in humans (Cooper, 1975) or pigeons (Hollard & Delius, 1982) also contained complex contours.

These contrasting results on rotational tests raise the question of how the stimulus information available in the two types of movement patterns might be encoded and stored. Because computational theories of motion perception have emphasized the possibility of extracting information about object structure from movement patterns, it might be feasible that an imaginary outline traced by the moving dot (resembling memorized visual persistance of the figure-8-type of configuration, for instance) could have been stored. This type of engram seems unlikely, however, because it would correspond closely to one of the outline stimuli used so that similar rather than discrepant results would have been expected on the rotational tests, in spite of the physical dissimilarity in the types of movement patterns.

Arguing along similar lines, if the engrams for moving-dot and moving-outline stimuli were stored in equivalent forms, then some consistent transfer effects would have been expected when birds were shifted from the discrimination of dot to similar contour stimuli. One of the stimuli in the initial training experiment was a dot trajectory produced by a 1 to 2 frequency ratio. In the second experiment, the same animals were trained on outline stimuli, one of which was generated by a 2 to 1 frequency ratio. The two tasks therefore included a transfer from a moving-dot stimulus in one axis orientation to an equivalent outline one, rotated through 90° (cf. Fig. 4.1 and 4.5). Conditions of transfer were balanced among subjects, with some birds having these potentially similar stimuli as $S+$ (or else $S-$) in both experiments. Such a positive transfer should have facilitated discrimination and led to higher discrimination accuracy at the beginning of the second experiment. For other animals, the transfer was negative. The similar configurations were $S-$ and $S+$ (or vice versa) in the dot and outline tasks. In this case initial performance following transfer should have been poorer. No differences in the predicted directions were obtained, however. It remains to be seen whether transfer effects could be obtained at all between corresponding dot and outline stimuli if birds were first primed to this correspondence by repeatedly shifting reinforcement contingencies between the two types of movement stimuli before testing on novel transfer patterns. However, the lack of spontaneous transfer between the two types of motion patterns, together with the discrepancies on the rotational tests, implies that, although the processing of both dot and outline stimuli must entail temporal integration, encoding of the movement patterns may differ.

It is tempting to speculate that present computational theories of motion perception deriving three-dimensional structure from a two-dimensional image (Ullman, 1984) nevertheless closely apply not only to humans' perception of movement patterns but also to pigeons'. In this case information of a higher order level about inferred object shape should be stored when moving-outline stimuli are discriminated, with motion parameters themselves then assuming secondary importance in this form of representation. According to these computational theories, three or four dots in conjugated motion, rather than a single dot as used here, are necessary to yield unambiguous information from which an object's shape can be derived. When only a single moving dot was presented, the pigeons may have memorized a movement pattern per se, devoid of inferred contours, rather than a representation of spatial structure.

Others have considered how these computational theories of motion perception may be implemented. Ballard (1986), for instance, has proposed the types of detectors that may underly movement-pattern perception, postulating that patterns consisting of contours and those composed of moving dots may be differently processed. Specifically, contours are said to stimulate oriented edge detectors, whereas the directional movement of dots stimulates blur detectors. Whether such computationally derived detectors have their counterparts in neural units in the avian (or mammalian) visual system remains to be seen.

What Are the Essential Parameters of Motion-Pattern Perception?

Although the pigeons easily learned to distinguish the various Lissajous figures described previously, it is still not clear what the essential parameters underlying this movement-pattern perception might be. In addition to simple direction of motion discussed earlier, a number of possible factors can be ruled out.

All the Lissajous figures that the pigeons had so far learned to discriminate were generated with sinusoids, the ratios of which were slightly offset from whole numbers. One frequency was approximately 1, 2, or 3 times the frequency of the other. The Lissajous figures thus produced consisted of trajectories or contour configurations that underwent a smooth and gradual change. It is conceivable that these particular frequency ratios are highly discriminable. An attempt was made to train birds that had previously had no difficulty in discriminating moving-outline figures with frequency ratios of 2 to 1 and 3 to 1 (see Fig. 4.5) to differentiate comparable dot-trajectory figures. The dot configurations were generated with similar ratios (1 to 2 and 1 to 3) but at lower frequencies. In spite of having had considerable experience with the discrimination of movement patterns in the "go/no go" task, the pigeons failed to reach a learning criterion within 20 sessions (Fig. 4.9). To human observers, these more complex dot tracks were also not as easy to differentiate as those used previously. In both

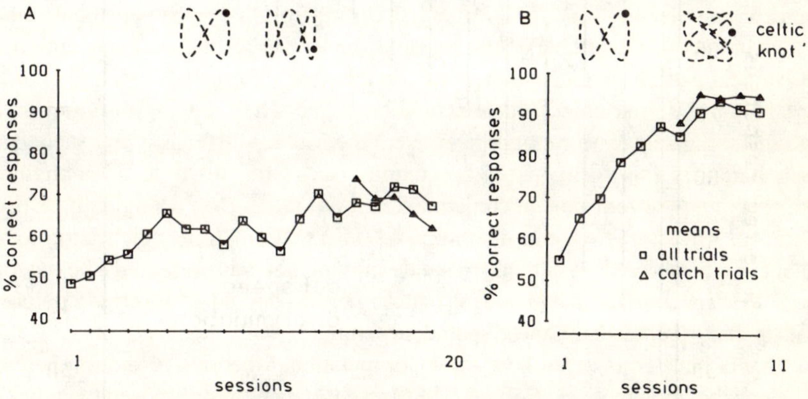

FIG. 4.9. Comparison of learning with different combinations of moving-dot stimuli. (a) Discrimination of two stimuli was poor when their variable trajectories both showed looping movement and underwent smooth transition. (b) Acquisition was swift when one stimulus had a smooth, looping movement and the other appeared more erratic in its motion. Momentary trajectories indicate the type of motion patterns.

stimuli, a dot was seen to loop around on the screen, one configuration merely appearing to be bouncier than the other.

The same pigeons were then transferred to another combination of moving-dot patterns, one of which again had a 1 to 2 frequency ratio. Its trajectory therefore underwent smooth transition. The ratio offset of the other stimulus (1 to 0.8) was such that, to human observers at least, the dot appeared to veer around eccentrically (Fig. 4.9). This stimulus was labeled *celtic knot*. Although its pathway was more difficult to describe, its motion was nevertheless easy for humans to distinguish from that of the smoothly varying dot pathway. Pigeons, too, very quickly learned to discriminate these stimuli. This suggests that, with dot motion at least, ease of discrimination depends, both for humans and pigeons, on global impressions of movement pattern. As the previous failure to transfer discrimination from outline to dot figures generated from similar frequency ratios shows, the numerical values of frequency ratios are not in themselves sufficient to predict whether figures will be readily discriminable.

Another parameter that was considered was stimulus velocity. When specifically trained to do so, pigeons can discriminate different velocities (Fig. 4.10; see also Mulvanny, 1978). This was tested by driving a single dot around a

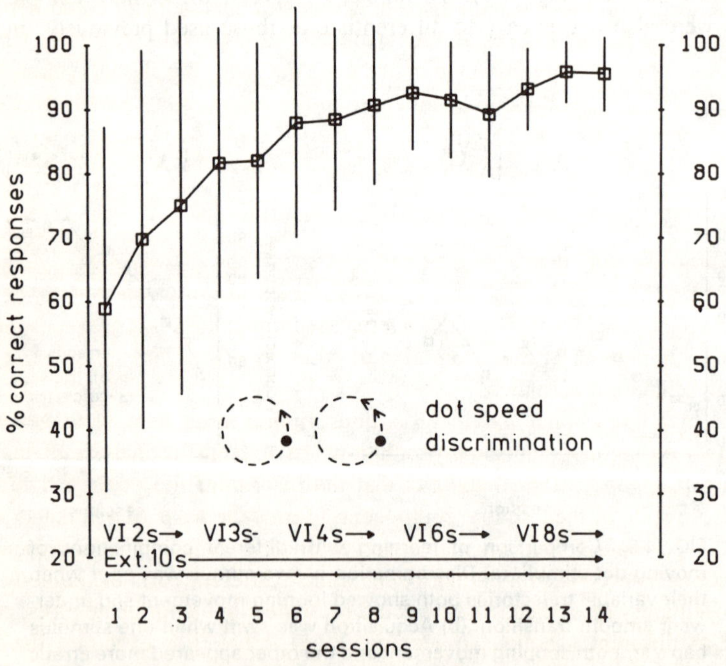

FIG. 4.10. Velocity discrimination of a moving dot. The dot described a constant circular track. Its angular velocity was either 360 or 720° per second.

constant circular pathway at an angular velocity of 360° per second or 720° per second. With the complex Lissajous figures, the dot's trajectory and its velocity varied from moment to moment. A priori it is not obvious what velocity parameter a pigeon might extract from the display, for example, the maximum velocity reached during some point in the cycle or an estimate of mean speed over a part of or the whole cycle. The outline figures consisted of gradually distorting contours in continuous motion. Estimations of velocity fields along such curves require complex calculations (Hildreth, 1984). But control tests indicated that stimulus velocity was not a sufficient discriminative cue.

By very slightly adjusting one frequency setting, the rate of phase shift (i.e., the time it took to complete a stimulus cycle) could be modified without altering overall stimulus configuration. With outline figures this could be used to produce a relative slowing down or speeding up in the contour drift and rate of figural change of either $S+$ or $S-$ stimulus. These changes in speed did not disrupt the pigeons' performance at all. Similarly, slight modifications in cycle rate did not affect discrimination of the dot stimuli. Furthermore, response latencies to the correct stimuli averaged about 1 second, irrespective of the time it took to complete a movement-pattern cycle. The cycle durations varied between 2 and 12 seconds for the different types of stimuli in these experiments. Results on the rotation tests with these stimuli also show that stimulus velocity could not have been a critical factor. Rotated figures included the same velocity components as nonrotated ones. If stimulus velocity had been decisive, the animals should have maintained a high level of discrimination on the rotation tests. With the exception of one bird, their discrimination accuracy instead decreased.

Lissajous Figures and Complex Motion Analysis

It was argued earlier that there are good ecological reasons why not only humans but also pigeons, and indeed most other visually guided animals, should be able to distinguish complex movement patterns. Motion-pattern perception should be especially important for the recognition of other organisms and the interpretation of their behavioral activities.

The two-dimensional moving Lissajous figures used here were generated from two orthogonal sinusoidal oscillations. A movement analysis of humans walking (Cappozzo, 1984) suggests that the discrimination of these Lissajous figures, in spite of their simple components, forms the basis of a human's or an animal's perceptual task when it observes another organism's movements in three dimensions. When a human being walks, different parts of the body perform oscillatory movements that, when represented on bicoordinate plots, correspond to different Lissajous trajectories (Fig. 4.11). For example, the head makes an oscillation of one cycle per footstep in the sagittal plane. This roughly corresponds to the elliptical pathway of a Lissajous figure with a 1 to 1 frequency ratio. At the same time the head moves through one cycle per two steps in the

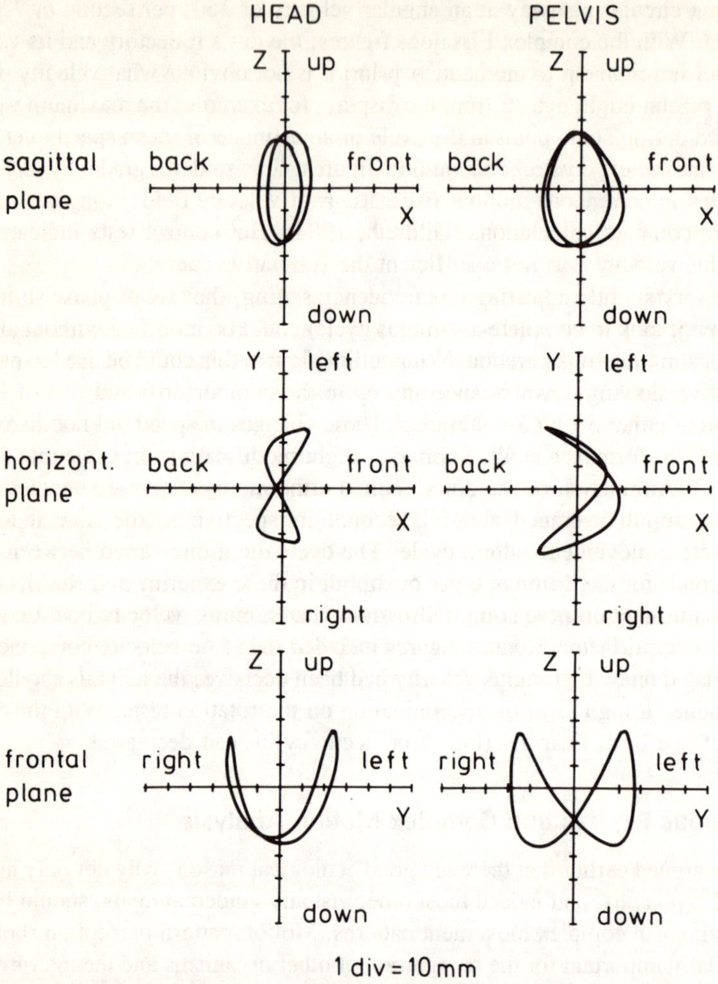

FIG. 4.11. Movement analysis of different parts of the body of a human walking. Component movement is referred to translational motion of the whole body (adapted from Cappozzo, 1984).

frontal plane. This oscillation approximates the U-shaped trajectory of a Lissajous figure with a 1 to 2 frequency ratio. The movement of other parts of the body could be similarly analyzed. This analysis suggests that the perceptual recognition of body movement consists in the parallel processing in different planes of component movements, each of which is describable as a Lissajous figure. This perceptual ability could be a fundamental property of all highly developed visual systems. Rhythmic motion patterns, themselves controlled by pattern generators in the central nervous system (Grillner & Wallén, 1985), have evolved to be characteristic of different organisms. Perceptual processes should be under strong selective pressures adapting them to recognize cyclic movements if an animal is to gain maximum information from its visual environment. The abilities of pigeons to discriminate complex motion might even be underestimated in the experiments described here. The Lissajous figures, presented on a pecking key, were probably viewed in the frontal visual field. Pigeons are known to be more sensitive to motion in the lateral than in the frontal field (Martinoya, Rivaud, & Bloch, 1983). However, the relation of threshold movement sensitivity to complex pattern recognition is far from clear (Nakayama, 1985).

The principle by which moving Lissajous figures are generated suggests a procedure by which any complex motion may be synthesized or, conversely, analyzed. Any motion of a point on a two-dimensional plane can be decomposed into the displacements of that point projected onto two orthogonal axes (Fig. 4.12). The displacements on the axes describe two continuous temporal functions. With Fourier analysis, any such continuous functions can themselves be reduced to sums of sine waves of specific frequency, amplitude, and phase. This procedure can obviously be extended to movement in three dimensions. Fourier-like procedures operating in the central nervous system seem to underlie both visual information processing (Kulikowski & Bishop, 1981) and the control of motor output (Pribram, Sharafat, & Beekman, 1984), which implies a common basis for the perception and generation of movement patterns. The periodicity of

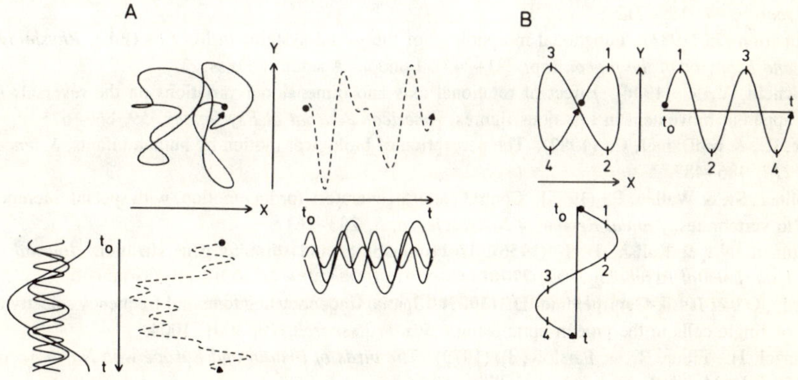

FIG. 4.12. (a) Complex motion of a point in two-dimensional space. The movement pathway here is generated from two complex temporal functions (dashed lines with amplitudes X and Y). These functions can be analyzed into their component sinusoids (superimposed continuous lines); (b) A simple Lissajous figure, generated by two sine waves. 1–4 are temporally corresponding points on X and Y axes. The complex trajectory (a) is obtained in a similar way.

biological motion could permit an animal to predict, at least in the short term, the future actions and spatial position of another organism and to adjust its own behavior accordingly. The study of the principles by which Lissajous figures are discriminated by pigeons could be fundamental to an understanding of the perception of biological action and of the complex motion of objects in space.

ACKNOWLEDGMENTS

This research was supported by the Deutsche Forschungsgemeinschaft through its Sonderforschungsbereich 114. I thank Prof. Juan D. Delius for numerous helpful discussions and Frank Münchow-Pohl, who provided technical enlightenment as well as an electronic circuit for the velocity- and direction-discrimination experiments. Ulli Schall helped to produce the spaghetti, and Annette Lohmann assisted with the preparation of the figures.

REFERENCES

Ballard, D. H. (1986). Cortical connections and parallel processing: Structure and function. *Behavioral and Brain Sciences, 9,* 67–120.

Cappozzo, A. (1984). Gait analysis methodology. *Human Movement Science, 3,* 27–50.

Cooper, L. A. (1975). Mental rotation of random two-dimensional shapes. *Cognitive Psychology, 7,* 20–43.

Cutting, J. E., & Kozlowski, L. T. (1977). Recognizing friends by their walk: Gait perception without familiarity cues. *Bulletin of the Psychonomic Society, 9,* 353–356.

Delius, J. D., & Nowak, B. (1982). Visual symmetry recognition by pigeons. *Psychological Research, 44,* 199–212.

Emmerton, J. (1983). Functional morphology of the visual system. In M. Abs (Ed.), *Physiology and behavior of the pigeon* (pp. 221–244). London: Academic Press.

Fisichelli, V. R. (1946). Effect of rotational axis and dimensional variations on the reversals of apparent movement in Lissajous figures. *American Journal of Psychology, 59,* 669–675.

Fox, R., & McDaniel, C. (1982). The perception of biological motion by human infants. *Science, 218,* 486–487.

Grillner, S., & Wallén, P. (1985). Central pattern generators for locomotion, with special reference to vertebrates. *Annual Review of Neuroscience, 8,* 233–261.

Guttman, N., & Kalish, H. I. (1956). Discriminability and stimulus generalization. *Journal of Experimental Psychology, 51,* 79–88.

Hardy, O., & Jassik-Gerschenfeld, D. (1979). Spatial frequency and temporal frequency selectivity of single cells in the pigeon optic tectum. *Vision Research, 19,* 1001–1004.

Heinzel, H., Fitter, R., & Parslow, J. (1972). *The birds of Britain and Europe with North Africa and the Middle East.*. London: Collins.

Hildreth, E. C. (1984). Computations underlying the measurement of visual motion. *Artificial Intelligence, 23,* 309–354.

Hodos, W., Smith, L., & Bonbright, J. C. Jr. (1976). Detection of the velocity of movement of visual stimuli by pigeons. *Journal of the Experimental Analysis of Behavior, 25,* 143–156.

Hollard, V. D., & Delius, J. D. (1982). Rotational invariance in visual pattern recognition by pigeons and humans. *Science, 218,* 804–806.

Johansson, G. (1973). Visual perception of biological motion and a model for its analysis. *Perception & Psychophysics, 14,* 201–211.

Kozlowski, L. T., & Cutting, J. E. (1977). Recognizing the sex of a walker from a dynamic point-light display. *Perception & Psychophysics, 21,* 575–580.

Kulikowski, J. J., & Bishop, P. O. (1981). Fourier analysis and spatial representation in the visual cortex. *Experientia, 37,* 160–162.

Lombardi, C. M., & Delius, J. D. (in preparation). *Pattern recognition in pigeons: Outlines, contrast, and color.*

Lorenz, K. Z. (1958). The evolution of behavior. *Scientific American, 199*(6), 67–78.

Martinoya, C., Rivaud, S., & Bloch, S. (1983). Comparing frontal and lateral viewing in the pigeon. II. Velocity thresholds for movement discrimination. *Behavioural Brain Research, 8,* 375–385.

Mulvanny, P. (1978). Velocity discrimination by pigeons. *Vision Research, 18,* 531–536.

Nakayama, K. (1985). Biological image motion processing: A review. *Vision Research, 25,* 625–660.

Philip, B. R., & Fisichelli, V. R. (1945). Effect of speed of rotation and complexity of pattern on the reversals of apparent movement in Lissajous figures. *American Journal of Psychology, 58,* 530–539.

Pribram, K. H., Sharafat, A., & Beekman, G. J. (1984). Frequency encoding in motor systems. In H. T. A. Whiting (Ed.), *Human motor actions—Bernstein reassessed* (pp. 121–156). North Holland: Elsevier.

Robbins, C. S., Bruun, B., & Zim, H. S. (1966). *Birds of North America.* New York: Golden Press.

Siegel, R. K. (1970). Apparent movement detection in the pigeon. *Journal of the Experimental Analysis of Behavior, 14,* 93–97.

Siegel, R. K. (1971). Apparent movement and real movement detection in the pigeon: Stimulus generalization. *Journal of the Experimental Analysis of Behavior, 16,* 189–192.

Sumi, S. (1984). Upside-down presentation of the Johansson moving light-spot pattern. *Perception, 13,* 283–286.

Tolhurst, D. J. (1973). Separate channels for the analysis of the shape and the movement of a moving visual stimulus. *Journal of Physiology, 231,* 385–402.

Ullman, S. (1984). Maximizing rigidity: The incremental recovery of 3-D structure from rigid and non-rigid motion. *Perception, 13,* 255–274.

5 Pattern Recognition, Updating, and Filial Imprinting in the Domestic Chicken (*Gallus gallus*)

Catriona M. E. Ryan
Stephen E. G. Lea
University of Exeter, UK

Much of this book is concerned with how birds categorize complex visual stimuli. This chapter is concerned with a particular kind of complex stimulus, namely another individual of the same species as the subject. An individual may be seen from many different points of view, and its appearance may also change as a function of season or maturation. It therefore clearly represents a polymorphous stimulus. Furthermore, the recognition of individual conspecifics is likely to be important in the social ethology of many species.

Most of the experiments described here used the slide discrimination techniques made popular by Herrnstein and Loveland's pioneering experiments with pigeons (Herrnstein & Loveland, 1964). The species under investigation, however, was not the pigeon, but the domestic chicken. Our aim in using chickens was not merely to extend slide discrimination to another species, but arose from a consideration of the biological importance of individual recognition; as we discuss later, free-range chicks grow up in family groups in which all the members appear to respond selectively to one another. There is abundant evidence that physical features such as those present in color slides are very important to chickens in the identification of other individuals (see, for example, Guhl & Ortman, 1953; Ramsay, 1951). Pigeons, on the other hand, may be less dependent on such features; it is known that for their near relatives, the ring-doves, behavioral cues are more important (Bennett, 1939). Furthermore, it has been suggested that young domestic chicks build up a composite image of their parents and siblings by the integration of a series of discrete images, or "snapshots," seen from different angles (Bateson, 1973). If this is so (and admittedly the evidence for it remains sketchy), it suggests that chickens should have few problems with the type of slide discrimination usually employed in the "concept discrimination" literature.

In using slide discriminations to investigate ecologically relevant discriminations, we are of course limiting the conclusions we can draw. We are not able to conclude that chickens *cannot* make certain kinds of discriminations—only that they cannot do so on the basis of the information included in color slides, produced by a technology adapted to the human, not the chicken eye. But we are able to demonstrate that some kinds of discrimination are readily learned. If others are learned more slowly, or not at all, that will be at least suggestive. This general strategy has of course been used with considerable success before, e.g., by Pietrewicz & Kamil (1981).

Ryan (1982a) demonstrated that cockerels rapidly learned to discriminate between slides on the basis of the individual hens depicted on them. Because this chapter focuses on the individual recognition that occurs between hens and chicks, it begins with a report of an experiment that replicated this result using hens as both subjects and objects. Two hens (the *object birds*) were photographed from several different angles, and the resulting prints were cut out and rephotographed against plain white backgrounds as slides. Twenty slides of one object bird comprised one stimulus set, and the second consisted of 20 slides of the other object bird. Eight further hens (the *subject birds*) were trained to discriminate between these sets. The procedure was very close to that described by Ryan (1982a), but a brief résumé is given here, as it was broadly similar for all our slide-discrimination experiments.

The subjects were given a series of training sessions, each comprising 240 trials, half of which were positive. The slides were back projected at life size from a Kodak "Carousel" random-access projector onto a 61.5 by 32.5 cm screen, 75 cm in front of the pecking key. The latter was in the form of a 6 by 6 cm window which the bird could look through as well as peck. (This was different from the procedure used by Ryan, 1982a, and most other experimenters in concept discrimination, in which a small image was projected onto a translucent key). For half the subjects, the positive slides were those in the odd-numbered slots in the projector magazine; for the remainder, they were those in the even-numbered slots. Each session involved 6 blocks of 40 trials; the 40 slides were presented in a different pseudorandom order in each block. A fixed-interval training schedule was used. For the first 30 seconds of a slide's presentation, pecks to the key were recorded but not reinforced. If the slide was positive, the first subsequent peck allowed 2.5 seconds access to a food-hopper, followed by an 8-second blackout: If it was negative, a 2.5-second and 8-second blackout followed immediately after the 30-second schedule interval had expired. Discrimination was measured by carrying out a Mann–Whitney *U*-test between the positive and negative slide groups, using the mean response rates to each slide as data. The birds were trained until they had attained a criterion of three consecutive sessions in which the *U* was significant at the 0.001 level or better, corresponding to a value of 0.78 of the rho statistic used by Herrnstein and his collaborators.

Figure 5.1 shows that all birds rapidly learned the discrimination and they subsequently showed good transfer to new slides of the training birds.

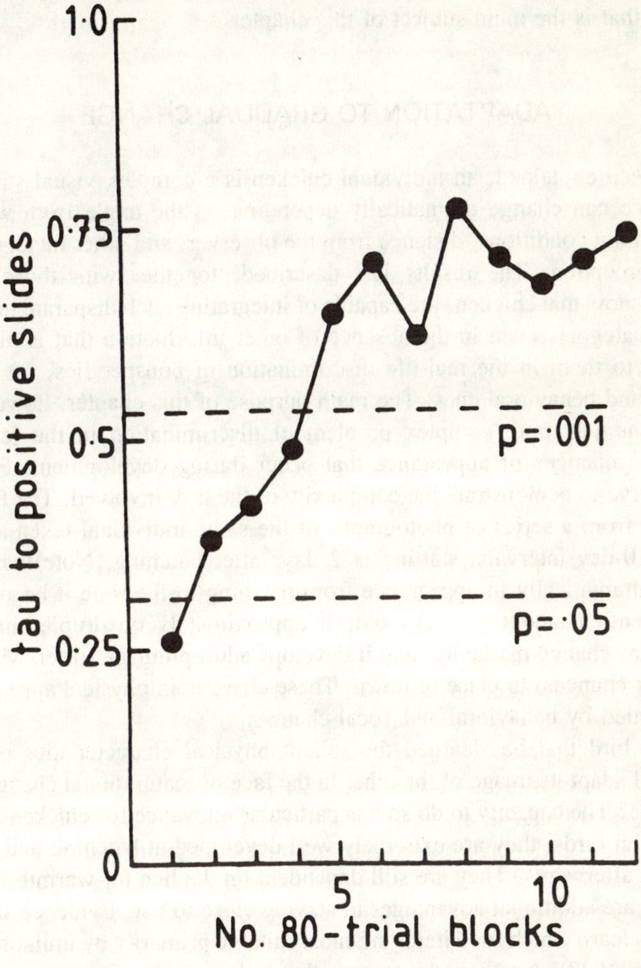

FIG. 5.1. Mean acquisition by eight subjects of a discrimination between two object birds. Twenty slides of each object bird were used. Discrimination (vertical axis) is expressed in Kendall's tau, which gives the correlation between the response rate to a slide and the object bird it depicts. Perfect discrimination would thus have a tau of 1 or −1, and random responding a tau of 0. The broken horizontal lines represent tau values where discrimination is significant at the 0.05 and 0.001 levels, respectively.

Tests for artifacts failed to reveal anything except the pictorial content of the slides that could account for the birds' performance. It therefore appeared that hens, like cockerels, are capable of learning slide discriminations of the type used with pigeons, when the categories to be discriminated are pictures of distinct individuals. This made it possible to use these techniques to investigate the problem that is the main subject of this chapter.

ADAPTATION TO GRADUAL CHANGE

As has been explained, an individual chicken is a complex visual stimulus; its appearance can change dramatically depending on the angle from which it is seen, lighting conditions, distance from the observer, and other factors affecting visual perception. The results just described, together with those of Ryan (1982a), show that chickens are capable of integrating such disparate images into a single category, even in the absence of other information that is presumably available to them in the real-life discrimination of conspecifics, for example, auditory and behavioral cues. The main purpose of this chapter, however, is to look at the yet more complex problem of discrimination in the face of the systematic changes of appearance that occur during development. Figure 5.2 should serve to demonstrate the complexity of the task involved. The figure was produced from a series of photographs of the same individual taken at approximately 10-day intervals, starting at 2 days after hatching. Note that the bird changes dramatically in appearance from hatching to the time it becomes fully independent, at about 6 weeks old. It approximately quadruples in size, its proportions change markedly, and it develops adult plumage (often with associated color changes) in place of down. These changes in physical appearance are accompanied by behavioral and vocal changes.

Can a bird that has learned the salient physical characteristics of another individual adapt its image of the other in the face of maturational changes of this magnitude? The capacity to do so has particular relevance for chickens because, as precocial birds, they are extremely well developed at hatching and leave the nest soon afterwards. They are still dependent on the hen for warmth, however, and there are additional advantages in staying close to her. Evidence shows that the chicks learn which food items are most suitable primarily by imitating the hen (Bruckner, 1933). Another advantage is that a chick that is a member of a family group is presumably less likely to fall victim to predation, both because of protection from the hen, and because animals in groups tend to be less vulnerable (Hamilton, 1971; Kenward, 1978).

If family unity is to be maintained, the hen and her chicks must be able to identify one another. As the previous experiment indicated, recognition of the hen is eminently possible—even in the unlikely event of the chicks relying solely on physical (as opposed to behavioral) features to do so. In addition, there is only one

FIG. 5.2. Stages in the development of a typical chick from 2 to 65 days old. The pictures are tracings of some of the original photographs used in our experiments. They are not drawn to scale, but the line under each is equivalent to 5 cm. Notice that the bird at 65 days old is approximately 4 times as large as it is at 2 days old. In addition, it acquires adult plumage (represented by continuous outlines) in place of down (broken outlines) and undergoes marked color changes as shown.

hen for the chicks to learn about. The hen, on the other hand, must not only learn about a brood of anything up to 12 chicks but must take account of the changes each undergoes as it matures. Given this asymmetry, it is reasonable to ask whether, in fact, the hen does learn about her chicks, or whether it is necessary for her to do so. Young chicks' social preferences are formed at a very early age (within 2 days of hatching) when they learn about their parent by the process known as imprinting (see later). Imprinted chicks tend to follow the parent and avoid strange moving objects. It is possible, therefore, that the strength of these preferences on their own might be enough to maintain family unity. On close examination, however, the effects of imprinting alone do not appear to be sufficient to keep the family together for the appropriate length of time. Bruckner

(1933) made extensive studies of hen–chick groups from the time the chicks hatched until they became fully independent at approximately 8 weeks old. He found that, although the group initially remained very close-knit with the chicks closely followed the hen, from as young as 10–12 days old the chicks began to investigate new stimuli and to forage further away from the hen. If danger threatened, however, the group would quickly reunite and remain "on alert" for some time.

Although the evidence is equivocal (Ramsay, 1951), we believe that this process depends partly on the hen's recognition of the growing chicks. Attempts to introduce strange chicks into an established brood are not normally successful, though the chances of acceptance are increased if the newcomers are similar in appearance and behavior to the familiar brood (Cushing & Ramsay, 1949). Cushing and Ramsay also found that the older the brood the more likely were strangers (even of a similar difference in age) to be rejected. Thus if the established brood was 10 days old and some day-old chicks were introduced, they would be accepted, but if the familiar chicks were 28 days old, strange birds 10 days younger would be rejected. Some rather similar (and equally ambiguous) results had earlier led Bruckner (1933) to suggest that, whereas a hen might be able to distinguish her chicks en masse from strangers, she could not identify each individually. He suggested instead that she had a "complex recognition impression" of the whole brood, which was sustained by the chicks' ability to identify her. Although this lays the major onus for familial recognition on the chicks, it does give the hen some part to play.

Observations of hen–chick groups in our laboratory indicate that the hens we use readily distinguish their own chicks from strangers. For example, we carried out a brief experiment with a "family" group where a broody hen had been given four artificially incubated eggs 24 hours before they were due to hatch. When the resulting chicks were 20 days old, one was removed and a strange chick of identical strain and age was placed in the run in a small cage. The hen attacked it without hesitation and made repeated attempts to peck it through the wire of the cage. When her own chick was replaced (in the same small cage), however, she made no attempt to peck it but instead stood some way from the cage and called the chick, which made strenuous efforts to escape and join her.

In the light of our slide-discrimination results, it would not be surprising if the hen's recognition of her chicks does play a part in family cohesion. If chickens have a highly developed ability to distinguish visually between other adult chickens, it is likely that they can also distinguish their developing young from strange chicks. But there is other evidence that chicks are identified throughout development, and it comes from the chicks themselves. It is well known that incubator-hatched chicks, reared without a parent, will imprint on one another (Guiton, 1959). According to our observations, chick "families" reared in this way do not differ noticeably from natural hen–chick groups in their familial behavior. In particular, these groups maintain their integrity, even though they lack the comparatively stable reference point provided by a mother hen. It is quite possible that imprinting on siblings may play an important role in maintaining family unity even when a parent is present. Furthermore, Bruckner (1933) found that chicks start to develop a peck-order from approximately 16 days of age and argues that once this is established the chicks must be said to "know" each other.

Nevertheless, arguments that chickens actually do maintain individual recognition across developmental changes do not remove the difficulties involved in achieving this, as Fig. 5.2 illustrates. It seems reasonable to ask how the parent (or sibling, for that matter) copes with such problems, and in the following experiments we set out to investigate this question.

The Developing Chick as a Natural Category

Although the task of keeping track of the changing appearance of a developing chick seems difficult to us, it does not logically follow that it will be difficult for a chicken. It may be, for example, that there is some subtle single cue present throughout development that serves as a basis for integration. Whereas it is not immediately obvious how a hen would know the feature(s) in a day-old chick on which to concentrate, the possibility of their existence cannot be dismissed out of hand on these grounds. Another possibility is that the hen and chicks generalize, using many stimulus dimensions, from a single fixed reference point (perhaps we should call it a prototype), established when family cohesion is at its maximum, when the chicks are under 12 days old. Our next experiment was designed to determine whether generalization from such a reference age would be sufficient to enable continued recognition, or whether birds need to update their images of developing chicks from time to time.

In this experiment, as in the last, subject birds were trained to discriminate between sets of 20 slides of 2 hen object birds, at a single age. But in this case, the ages of the object birds at the time the training slides were taken was different for different subjects. Twelve experimentally naive hens served as subjects. Four of these were trained to discriminate between slides of the object birds at 2 days old, 4 more to discriminate between slides of the same 2 object birds taken when they were 33 days old, and 4 learned to discriminate between slides taken of them when they were 65 days old, by which time their plumage was fully mature, though their combs were not fully developed. When each subject had reached the learning criterion, it was given a generalization test, in which the new slides were pictures of the same 2 object birds, but at all the 7 ages illustrated in Fig. 5.2, i.e., 2, 12, 23, 33, 43, 55, and 65 days. It was thus possible to estimate how far generalization to new ages was possible from a single reference point. The results are summarized in Fig. 5.3. Figure 3a shows the mean acquisition by the three group of subjects. Notice that all groups readily learned their discriminations and there was little difference in the rates at which

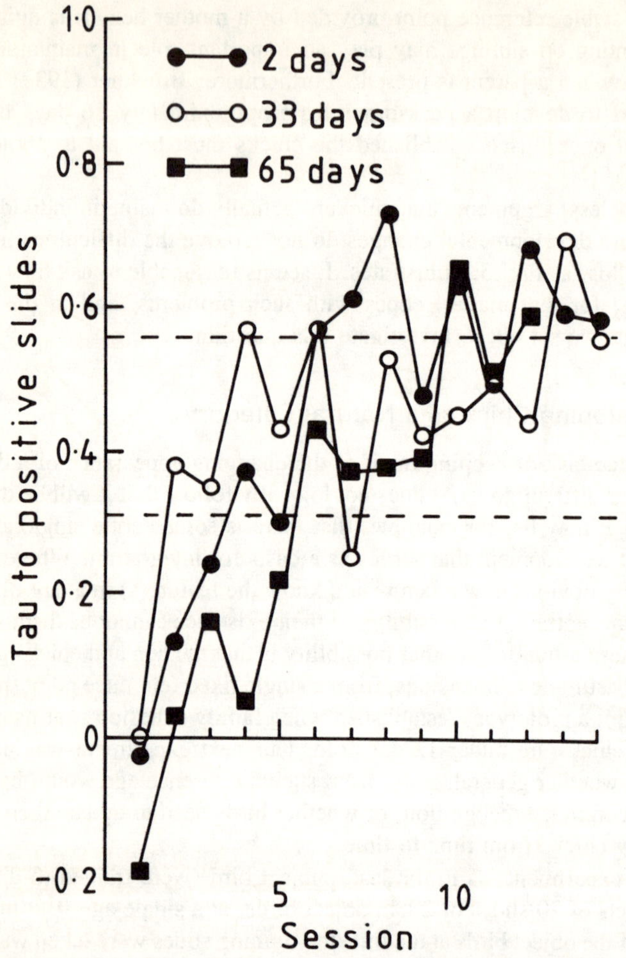

FIG. 5.3a. Acquisition of a discrimination between the same two object birds at 2, 33, and 65 days old. Each curve shows the mean performance of the 4 birds trained on that discrimination. Performance is measured in terms of Kendall's tau (see Fig. 5.1), and broken horizontal lines represent tau values at which discrimination was significant at the 0.05 level (lower line) and 0.001 level (upper line).

they did so (the apparently faster acquisition of the 33-day discrimination was due to a single subject in that group).

Figure 5.3b shows the mean proportion of responses to positive slides within ages during the transfer cycle (note that the slides at the training ages were not new but were some of the familiar training slides). As expected, all birds performed best with the slides on which they had been trained. What is interesting,

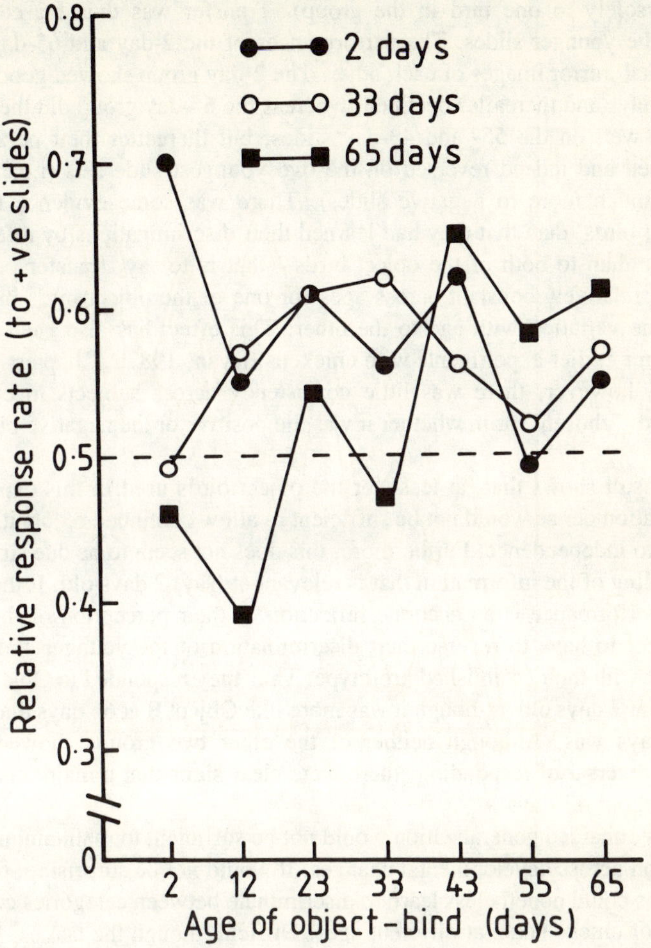

FIG. 5.3b. Mean proportion of responses to positive slides within ages during the transfer cycle of the same experiment. Response rates above 0.5 (broken line) thus indicate more responding to positive slides than to negative ones; those below 0.5 show the opposite. All slides of the object-birds at the training ages were taken from the slide sets used in training: the remaining slides were new to the subjects.

though, is that the generalization curves are not uniform, as would be expected if the birds were responding to a smooth and gradual change. Rather, there is some evidence of a cut-off point at around the 43-day level, after which the responses are markedly altered. Thus, birds trained on 33-day slides showed good transfer to the 23-day slides and some transfer to 12- and 43-day slides. Beyond this, their performance declined to chance level (the apparent improvement at 65 days was due solely to one bird in the group). Transfer was thus biased slightly towards the younger slides. The performances of the 2-day and 65-day groups were virtual mirror images of each other. The 2-day group showed good transfer up to 43 days and thereafter declined, whereas the 65-day group did the reverse: They did well on the 55- and 43-day slides, but thereafter their performance deteriorated and indeed reversed on the two youngest slide sets (i.e., they responded much more to negative slides). There was some evidence from the individual birds' data that they had learned their discriminations by reference to one rather than to both of the object birds—that is to say, transfer peck rates would be relatively constant across ages for one of the object birds but would show some variation with age to the other. This effect has also been noted in some of our earlier experiments with chickens (Ryan, 1982b, Chapters 3–4). In this case, however, there was little consistency across subjects in either the object bird "chosen" or in whether it was the positive or the negative bird of the pair.

This result shows that, at least for the object birds used in this experiment, generalization per se would not be sufficient to allow continued recognition from hatching to independence. Furthermore, this does not seem to be due simply to a steady fading of the information that is relevant at (say) 2 days old. If the 65-day group's performance is an accurate reflection of their perception of the slides, they appear to have to reverse their discrimination of the younger slides to be consistent with their established prototype. Thus they responded to Object A (for example) at 2 days old as though it was more like Object B at 65 days than Object B at 2 days was. Although neither of the other two groups showed such a dramatic reversal of responding, there were clear signs that a major change did occur at about 43 days.

Clearly, unaided generalization would not be sufficient to maintain individual recognition across developmental changes. It would not be surprising, however, if chickens could nonetheless learn to discriminate between categories consisting of slides of chicks taken at different ages. Difficult though the task is, it should not be more difficult than learning the kind of "random pseudoconcept" that has been used by Vaughan (cf. Vaughan & Greene, 1984; Vaughan, this volume; for discussion, see Lea & Ryan, this volume). In our next experiment, therefore, we exposed hens to a quite unnatural discrimination. Like the previous experiments, it involved learning a discrimination between two slide sets. Each set consisted of pictures of one of the 2 object birds used in the last experiment. But in this case each set included 5 pictures of the relevant object bird at each of the 7 ages shown in Fig. 5.2. Again 12 experimentally naive hens were used as subjects. Four were trained to discriminate the full sets of 35 slides. Two other groups, also each of 4 birds, were trained on subsets of 20 slides, covering the 2–33-day and 33–65-day age ranges, respectively. The criterion of learning was 3 consecutive sessions where discrimination was significant at the 0.05 level or better, by the usual Mann–Whitney test (corresponding to rho values of 0.61 for the 7-

age group, and 0.65 for the 4-age groups). At this stage each bird was given a transfer test in which the transfer stimuli consisted of new slides of the same birds at the familiar training ages, plus, for the 4-age groups, slides of the same birds at ages that the subjects had not previously seen. The results are summarized in Fig. 5.4.

Notice that acquisition (Fig. 5.4a) was markedly inferior to that in the first experiment. Indeed, 3 of the 12 subjects had failed to learn the discrimination after 30 sessions, at which point training was stopped. Furthermore, those subjects that did reach criterion did not discriminate between birds equally well for all age groups. Figure 5.4b shows the mean relative response rates to familiar positive slides within age groups during the transfer cycle and represents asymptotic performance on familiar slides. All subjects, whether they were learning the discrimination across the whole age range or subsets of it, performed best with "younger" slides and worst with the "oldest" sets. Thus, birds trained on the 2–33-day subset reached the learning criterion in a mean of 7.5 sessions fewer than those trained on the 33–65-day age group. This difference was just short of significance, but this was due to the performance of 1 bird in the 33–65-day group, which learned much faster than the others. The rest of this group did

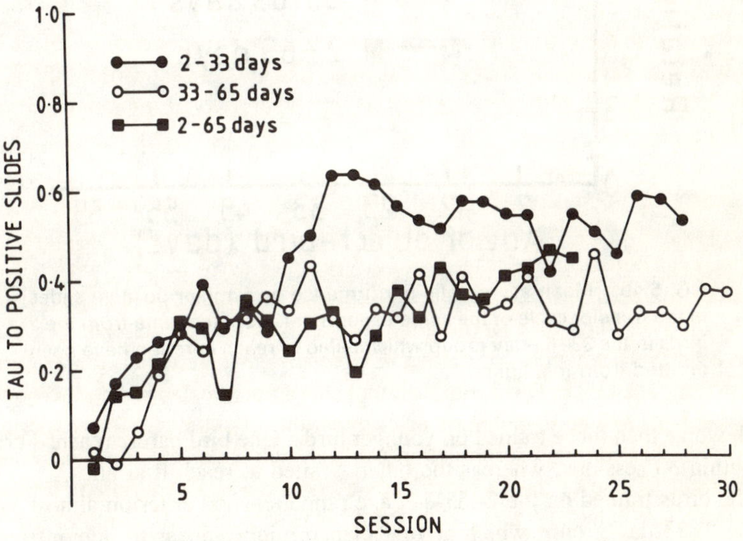

FIG. 5.4a. Acquisition of slide discrimination between 2 object-birds, where the slide sets contained pictures of those birds over a large (2–65 day) or small (2–33 and 33–65 day) age range. Each curve shows data from the 4 birds in the relevant group. All birds in the 2–33-day and 2–65-day groups successfully reached criterion within thirty sessions, but 2 birds in the 33–65 day group failed to do so.

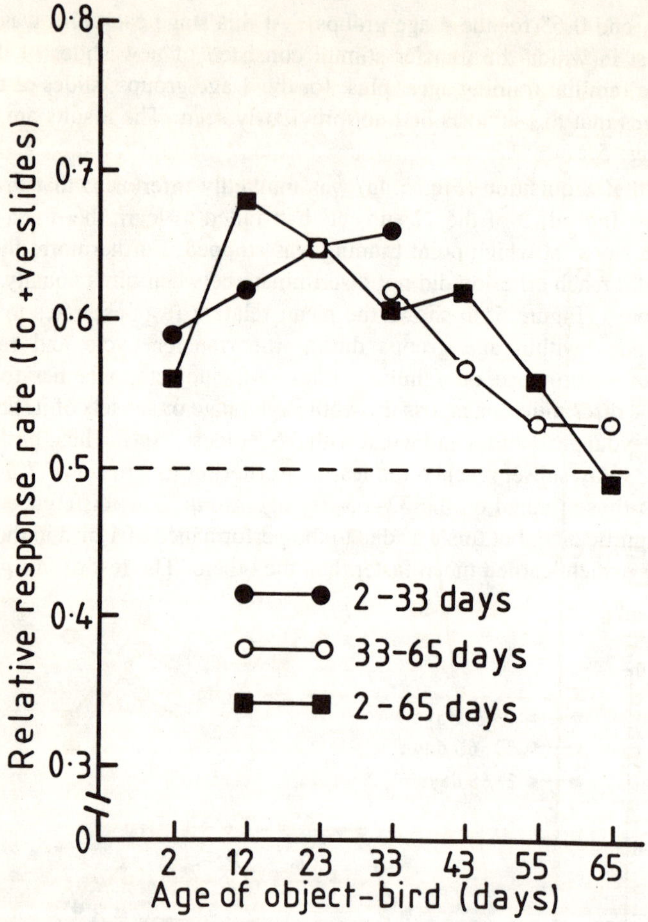

FIG. 5.4b. Mean asymptotic performance on familiar positive slides in the transfer cycle of the same experiment. Note that data from the 2 birds in the 33–65-day group which failed to reach criterion have been omitted from this figure.

much worse than those trained on younger birds: One bird barely reached criterion within 30 sessions, whereas the other 2 failed to reach it at all.

The birds trained on the 2–65-day age range reached criterion almost as fast as the 2–33-day group, which at first glance might appear to demonstrate an impressive capacity for categorization. Figure 5.4b, however, shows that their performance on "older" and "younger" slides was very similar to that of the birds trained on those ages alone. In other words, they were discriminating between the "younger" slides well enough to give them a significant overall performance, even though their discrimination of the "older" slides was barely above chance level. Their performance in the transfer test (Fig. 5.4c), however, suggested that they had learned something about the "older" slides, because their performance was marginally better than that of the 2–33-day group (which had not seen any "older" slides before). Similarly, they responded slightly more consistently than either of the subgroups to the "younger" slides, which suggests that they might have been better able to integrate all of them than were the

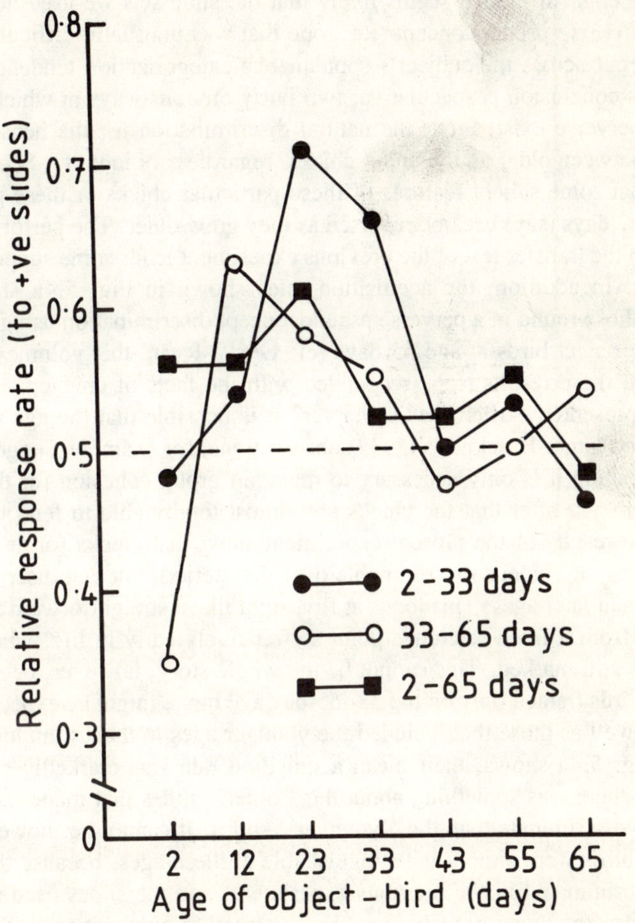

FIG. 5.4c. Mean relative response rates to positive slides within ages during the transfer cycle of the same experiment. Curves for the 2–33- and 2–65-day groups were derived from data from all 4 birds in the group: data from the 2 birds which failed to learn the discrimination were omitted from the curve for the 33–65 day group. Note that all slides (even those of the object birds at familiar ages) were new to the subjects.

latter birds. The number of subjects, though, was small, leaving the data vulnerable to the effects of individual differences in performance. For this reason, group differences of this nature must remain somewhat speculative.

The subjects clearly found the task difficult, and the contrast with the acquisition rates observed in the previous two experiments is sharp. It appears that a developing chick over the age range used here does not represent a coherent category for other chickens. In view of the ease with which pigeons learn random pseudoconcepts, in fact, it seems likely that the slide sets we used here represented a *perverse* pseudoconcept (i.e., one that was unusually difficult to learn because it cut across the subject's spontaneous categorization tendencies). Although this conclusion is speculative, two fairly obvious ways in which the sets could be perverse exist. First, the natural discrimination for the hens to make might be between older and younger chicks, regardless of identity. Second, it is possible that some salient features of these particular chicks or these particular pictures at 2 days (say) became reversed as they grew older. The performance of the birds in the transfer test of the previous experiment lends some support to this hypothesis. In addition, the acquisition rates shown in Fig. 5.4a are in fact similar to those found in a perverse pseudoconcept discrimination using slides of these same object birds at age 55 days (cf. Lea & Ryan, this volume).

How are these results to be reconciled with the facts of chicken social ethology, as presented earlier in this chapter? It is possible that the age range we used was too large. Bruckner's (1933) observations, for example, suggested that strictly speaking it is only necessary to maintain group cohesion for the first 6 weeks of life, as after that the chicks are almost totally able to fend for themselves. The results of the present experiment show that chicks (or at least our object chicks) are indeed discriminable over that period, but that discrimination broke down at later ages. This looks at first sight like a straightforward failure to generalize from a fixed reference point set relatively early in life, when group cohesion is still marked. This cannot be the whole story, however, because if it were, the birds trained only on the 33–65-day age range might be expected to do at least as well as those that included the younger ages in their stimulus sets. In fact, as Fig. 5.4a shows, their mean acquisition rate was markedly slower. It looks as if there was something about the "older" slides that made them more difficult to discriminate than the "younger" slides. It cannot be, however, that the object birds were simply indistinguishable at these ages, because they were readily discriminated in the previous experiment, and the slides used this time were a subset of those used before. The alternative explanation is that these pictures were ambiguous in the context of the age range as a whole, or perhaps in the context of some younger reference age.

It would be interesting to pursue this context effect. However, from the point of view of understanding the chicken's discrimination of conspecifics, there is a more important direction for further research. The task we set our chickens was in an important sense unnaturally difficult. Clearly, no bird is ever going to be exposed to the same chick at different ages simultaneously. A hen or chick will only have to cope with a relatively gradual change in appearance lasting several weeks. It may well be that this is a crucial difference, and the rest of this chapter is devoted to examining its implications.

UPDATING AND IMPRINTING

Logically, perhaps, the next step in this investigation should have been an experiment in which hens discriminated slides taken at different ages, introduced gradually in a developmental sequence. That experiment remains to be done, because we wished to use chicks rather than hens as subjects. As we pointed out earlier, the evidence that chicks can maintain discrimination of other chicks through developmental changes is somewhat stronger than it is for hens. This implied a change in experimental paradigm, from slide discrimination to imprinting—or, more precisely, *filial* imprinting, the process by which young nidifugous birds develop a following response towards objects seen shortly after hatching.

Since it was popularized by Lorenz (1935), the concept of imprinting has undergone many changes (for a review, see Bateson, 1966). Three of these revisions are particularly important for this chapter. First, it is no longer thought that any arbitrary stimulus will attract the following response that characterizes imprinting. For example, a moving and/or noisy stimulus is much more effective than a stationary and/or silent one, and, as might be expected, a live hen is more effective than an artificial stimulus (Boakes & Panter, 1985). Second, it is no longer believed that there is a strict *critical period* (18–36 hours post/hatching for chickens) during which exposure to a suitable object must occur if imprinting is to be observed; rather, this is viewed as a *sensitive period,* which is optimal for the occurrence of imprinting but does not exclude imprinting to objects seen at other times. Finally, and most important for our purposes, the effects of imprinting are no longer thought to be irreversible.

Normally, preferences formed during the sensitive period are extremely resistant to change (Schein, 1963). But if the initial object is comparatively weak as an imprinting stimulus (for example, if it does not move), the chick may subsequently come to prefer a more attractive stimulus introduced later (Boakes & Panter, 1985). There is also evidence that the natural sensitive period may be extended if the chicks are sensorily deprived from the moment of hatching (Moltz & Stettner, 1961). It is, however, extremely difficult to prevent some form of imprinting from taking place—for example, chicks reared in isolation, but in the light, have been reported to imprint on the walls of their cages (Bateson, 1964).

One of the main areas of research into the phenomenon of imprinting is the effect it has on preferences in later life. This has been studied with particular reference to *sexual imprinting,* which narrows a bird's sexual preferences to individuals similar to the birds with which it has been reared (Immelman, 1972). Normally, of course, these would be its parents or siblings, but experimental manipulation can produce some spectacular mismatches. For example, Schein (1963) imprinted turkey chicks on humans; when they became sexually mature, they courted humans rather than other turkeys. It is a vexed question whether sexual and filial imprinting are separate processes, and the evidence is conflicting. The sensitive period for sexual imprinting is much later than that for filial imprinting and occurs at around 6 weeks old in chickens (Vidal, 1976, as cited by Bateson, 1979). On the other hand, younger chicks that have been made sexually responsive by testosterone injections will direct their sexual behavior to the filial imprinting object (Bambridge, 1962). One way of resolving this conflict would be to appeal to the kind of updating process that we are using to account for the maintenance of individual recognition across development, and this lends further interest to an extension of the present work into the area of imprinting.

The major evidence for updating to maintain recognition of the imprinted stimulus, however, comes from observations of filial imprinting on siblings. Although this evidence is strong, much of it is either anecdotal or was gathered incidentally in the course of other investigations. We therefore carried out a brief pilot experiment to demonstrate imprinting on a changing stimulus. A group of 4 incubator-hatched chicks were reared from hatching for 50 days without any adults present, but with a chick 10 days older than they were. During the experiment, this chick underwent changes in appearance similar to the bird pictured in Fig. 5.2; furthermore, the experiment spanned the period from 33 to 55 days that seemed, from the slide discrimination experiments, to be some kind of cut-off point for stimulus generalization. Nonetheless, the younger chicks imprinted on the older one and remained so throughout the period of the experiment, with only the normal waning due to age. They continued to follow this chick even when its (virtually identical) siblings were introduced into the run.

Filial imprinting is a powerful mechanism for ensuring that all the chicks in a brood will remain together and in close association with the mother hen, for at least the first 2 weeks after hatching, simply because each has independently imprinted on her and on its siblings. The results reported thus far in this chapter strongly suggest that imprinting on siblings must involve some form of updating process if recognition is to be maintained throughout development. What is the nature of this process? Three possibilities need to be considered.

First, the chick might learn to follow the successive versions of its siblings only because the original imprinting stimulus (i.e., the sibling at less than 2 days old) is not available. This hypothesis would predict that if the chick could be given a choice between the original and updated versions of its imprinting stimulus (clearly an impossibility when this is a single sibling), it should prefer the original.

Second, the chick might periodically update its representation of its siblings, with each update overwriting all previous representations. In this case, when the chick is given a choice as previously mentioned, it should prefer the updated version. Some support for this view has been provided by Salzen and Meyer (1968), who found that chicks already imprinted on a particular object could nonetheless be induced to prefer a second object if they were exposed to it for a sufficient period of time. Such a mechanism is perfectly adequate for the short-term task of identifying siblings. In the longer term, however, there might be some selective advantage in a more complex process. If a hen retains some memory of her siblings from when they were very young, she might be better equipped to identify her own offspring (which should share at least some of their uncles' and aunts' traits).

Finally, the chick might gradually enlarge its category of "sibling" to incorporate the earliest imprinting stimulus and all subsequent changes to it. If this were so, the chick should show no preference if given a choice between original and updated versions of the imprinting stimulus.

The final experiment described here was designed to distinguish between these three possibilities. The imprinting stimulus used was a string of four table tennis balls, with a small bell attached to the bottom. Table tennis balls were chosen because they were cheap, readily available, and uniform units that could be used to build systematically variable stimuli. The bell, as a small bright object, reliably elicited pecking from the chicks. This in turn caused the bell to ring, which increased the effectiveness of the stimulus as a whole.

The experiment involved 121 chicks. For 59 of them, the imprinting stimulus consisted of a string of white balls; for the remainder the balls were sprayed with brown paint. Each chick was individually exposed to a moving string of balls for 30 minutes, 24 hours after hatching. It was then housed in visual isolation from other chicks, with the balls suspended in its cage. The birds were divided into two groups. In the first (the Experimental group) the balls on the string were changed, one at a time, for balls of the opposite color. Changes were made every 4 days; thus, from the 17th day onwards, all the balls in the string were of the opposite color to those of the original stimulus; birds imprinted on an all white string now had an all brown one, and vice versa. The second (the Control group) was divided into three subgroups. The birds in the first of these were exposed only to the original imprinting stimulus for the entire 21 days of the experiment. Those in the second subgroup were exposed to the imprinting stimulus for the first 17 days, after which a string of the opposite color was substituted. Finally, the chicks in the third subgroup were exposed to the imprinting stimulus for the first 4 days, after which it was removed and not replaced.

On the 21st day after hatching, all birds were given a preference test. Each was individually given 20 trials in a Y-maze, in which they could choose between the original imprinting stimulus and a string of balls of the opposite color. The preference for the imprinting stimulus was expressed as the proportion of the total number of trials in which the bird chose that stimulus. The results are shown in Fig. 5.5. There was some overall preference for brown as against white balls, so the results have to be analyzed taking this color preference into account.

Regardless of the original imprinting color, the Experimental group as a whole showed no significant preference between the test stimuli (two-tailed binomial tests). The median number of choices of the white balls by the chicks originally imprinted on white was 8, whereas for the originally brown-imprinted chicks it was 9. The numbers of choices of these two groups do not differ significantly, showing that the original stimulus had no effect on final choice. In contrast, birds in the first control condition showed a strong preference for the imprinting stimulus. The median number of choices of white balls was 15 for the white-imprinted chicks and 2 for the brown-imprinted chicks under this condition. This preference is significantly greater than that shown in the Experimental group (two-tailed Mann–Whitney test, U[32,10] = 1.5; $p < 0.001$ for brown-imprinted birds and 16.5; $p < 0.001$ for white-imprinted birds).

The second control condition, in which there was a sudden change to the opposite color at day 17, led to a weaker effect of original imprinting on final preference: the originally white-imprinted chicks made a median of 12 choices of the white balls, as compared to a median of 4 for the originally brown-imprinted chicks. The statistical difference between this condition and the Experimental group is once again highly significant, but only just so for the white-imprinted chicks (two-tailed Mann–Whitney test, U[32,10] = 71; $p < 0.01$, and 137; $p < 0.05$ for brown- and white-imprinted chicks, respectively). We can therefore say with confidence that gradual change has more effect on the originally imprinted preference than an abrupt change at 17 days.

The results of the third control condition are more variable, and a growing preference for white exists (at least among white-imprinted birds). The median number of choices of white balls by brown-imprinted birds was 8 (virtually the same as the Experimental group) but that by white-imprinted birds was 15, which is significantly different from the Experimental birds (two-tailed Mann–Whitney test, U[32,7] = 104.5; $p < 0.01$). Several chicks in this group, however, failed to approach either string of balls within 2 minutes in the Y-maze on a number of trials, so that null preferences were recorded. This suggests that although imprinting from days 2–5 had some effect on later preference (the white-imprinted chicks preferred white significantly more than the brown-imprinted chicks: U [10,7] = 2.5; $p < 0.001$), the bond had to some extent been weakened.

It was explained earlier that there are three possible ways in which a chick can remain imprinted on a developing sibling. The first of these is that it transfers its allegiance only in the absence of its preferred stimulus, which nevertheless retains its superior attraction. This would presumably be the line adopted in a strict view of the irreversible effects of imprinting. As we have seen, the earlier workers such as Schein (1963) had good grounds for believing that preferences formed during imprinting were, at the very least, extremely hard to change. Our results, however, are more consistent with the modern view that imprinting is not

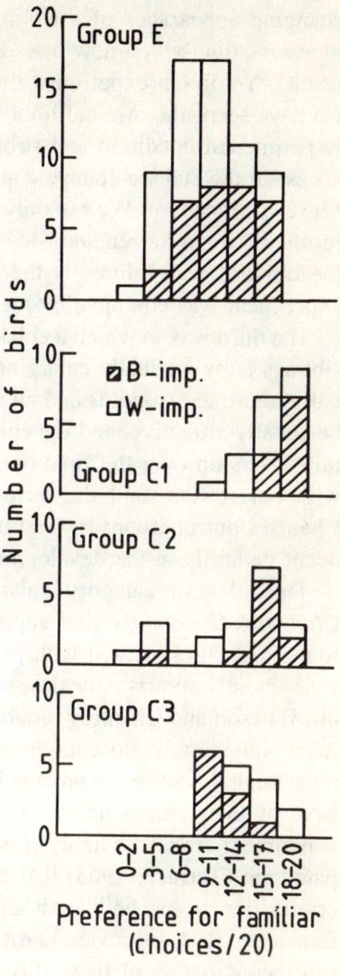

FIG. 5.5. Distribution of preferences for familiar vs. novel stimuli in the imprinting experiment. Chicks in Group E (N=64) had been exposed successively to all the intermediate stages between the imprinting (=familiar) and final (=novel) versions of the stimulus for the 21 days before the preference test. Chicks in Group C1 had only been exposed to the imprinting stimulus for the same period. Chicks in Group C2 saw the imprinting stimulus for the first 17 days and then the final version for the remaining 4 days. Birds in Group C3 saw the imprinting stimulus for the first 4 days, after which it was removed and not replaced.

entirely rigid; experience after the 6th day certainly had an important influence on subsequent preferences. This, is not, of course, a new finding. Salzen and Meyer (1968), for example, criticized Schein's work on the grounds that all his attempts to get his turkeys to reverse their preferences had been conducted in the presence of the original imprinting stimulus, namely a human being. Not surprisingly, the turkeys continued to prefer people to other turkeys. Salzen and Meyer, using chicks as subjects and green and blue balls as imprinting stimuli, obtained a clear transfer of preference from the original imprinting stimulus to a different stimulus substituted 3 days after hatching. This new preference was maintained even after 3 days deprivation of both stimuli.

The second way in which imprinted preferences could keep pace with the changing appearance of a sibling would be successive replacement of the old representation with a new one. This is sufficient to explain Salzen and Meyer's results. Yet in our experiment it would imply a strong recency effect. In fact, as we have seen, the original imprinting stimuli remained slightly preferred in our Experimental condition and strongly preferred (after allowing for color preferences) in the sudden-change control, which most closely resembled Salzen and Mayer's procedure. We can only speculate about the reasons for the difference in results between Salzen and Meyer's experiment and the present one, but among the obvious possibilities is the difference of time scale (Salzen and Meyer's experiment was complete by the 10th day after hatching).

The third way in which a chick can cope with the developmental changes in its siblings is by gradually enlarging its category of ''sibling'' to include the originally imprinted stimulus and all subsequent versions of it. Thus all versions will be equally attractive, and the chick should show no particular preference for the initial version over the final one. This is the possibility that is most consistent with our results, and the difference between the Experimental and Sudden-Change Control groups suggest that the gradual introduction of change, as would occur naturally in the development of a sibling, is necessary to its success.

This idea of category enlargement has something in common with the *Clumping Hypothesis* first suggested by Bateson and Chantrey (1972). They found that chicks reared with two different stimuli present simultaneously were subsequently surprisingly slow when trained to discriminate between those stimuli. Bateson and Chantrey proposed that this was because the birds had incorporated both stimuli into a single category. They also suggested that such a mechanism might underlie the process by which chicks integrate the first glimpses they have of their parent into a single category of *Parent*. The conditions of our experiment are also notably similar to the stimulus-fading or error-free learning paradigm (Terrace, 1963) that has been used with considerable success in discrimination tasks, both with animals (Terrace, 1966, 1972) and humans (see Lambert, 1980, for review). Among the tasks remaining for future research is a full reconciliation of these different ways of approaching what is essentially a single problem.

The capacity for category enlargement cannot be infinite, if useful discriminations are to be maintained. We know, for example, that chicks imprinted on a particular hen will avoid other hens of the same strain, even though the latter must have a great deal in common with their own mother (Bruckner, 1933). What would happen if, in a slide discrimination, we took subjects through a series of gradual changes that eventually reversed the original discrimination? Our guess is that the discrimination would be maintained, even though this would require the erasure of part of a category, rather than its extension, but this is another experiment that remains to be done.

At the beginning of this chapter, we acknowledged that we were setting aside possible nonvisual cues for individual recognition. This is not in any way to deny their role, which under natural conditions may well be crucial. It could even be that we have been wasting our time looking for evidence of updating, because nonvisual characteristics may provide sufficient continuity to enable continued recognition without it. What we have been aiming to do, however, is to see what kind of concept discrimination capacity chickens would need, if they were to maintain individuals recognition, despite developmental change, by static visual cues alone. Our results suggest that the discrimination of individuals at a single age is easy, but a category made up of all the ages of an individual is too complex to support easy discrimination. Nevertheless, it may be possible to expand the category of recognizable views of an individual, from a small, easily learned set to a large set that could not be learned all at once. This notion echoes the results of Cerella (1979) when he trained pigeons to discriminate between silhouettes of leaves. It suggests a number of further lines of investigation, which should be of interest from the point of view of both the study of concept discrimination and the social ethology of individual recognition.

ACKNOWLEDGMENTS

The research described in this chapter was supported by the United Kingdom Science and Engineering Research Council, through a research studentship to CMER. We would also like to thank Mr. John Redler, who supplied most of the subject birds for this research.

REFERENCES

Bambridge, R. (1962). Early experiences and sexual behaviour in the domestic chicken. *Science, 136,* 259–60.

Bateson, P. P. G. (1964). Effect of similarity between rearing and testing conditions on chicks' following and avoidance responses. *Journal of Comparative and Physiological Psychology, 57,* 100–103.

Bateson, P. P. G. (1966). The characteristics and content of imprinting. *Biological Reviews, 41,* 177–220.

Bateson, P. P. G. (1973). Internal influences on early learning in birds. In R. A. Hinde & J. Stevenson-Hinde (Eds.), *Constraints on learning* (pp. 101–116). London: Academic Press.

Bateson, P. P. G. (1979). How do sensitive periods arise and what are they for? *Animal Behaviour, 27,* 470–486.

Bateson, P. P. G., & Chantrey, D. F. (1972). Retardation of discrimination learning in monkeys and chicks previously exposed to both stimuli. *Nature, 237,* 173–174.

Bennett, M. A. (1939). The social hierarchy in ring doves. *Ecology, 20,* 337–357.

Boakes, R., & Panter, D. (1985). Secondary imprinting in the domestic chick blocked by previous exposure to a live hen. *Animal Behaviour, 33,* 353–365.

Bruckner, G. H. (1933). Untersuchungen zur Tiersoziologie, insbesondere zur Auflosung der Familie. *Zeitschrift für Psychologie, 128,* 1–115.

Cerella, J. (1979). Visual classes and natural categories in the pigeon. *Journal of Experimental Psychology: Human Perception and Performance, 5,* 68–77.

Cushing, J. E., & Ramsay, A. O. (1949). The non-heritable aspects of family unity in birds. *Condor, 51,* 82–87.

Guhl, A. M., & Ortman, L. L. (1953). Visual patterns in the recognition of individuals among chickens. *Condor, 55,* 287–298.

Guiton, P. (1959). Socialization and imprinting in Brown Leghorn chicks. *Animal Behaviour, 7,* 26–35.

Hamilton, W. D. (1971). Geometry for the selfish herd. *Journal of Theoretical Biology, 31,* 295–311.

Herrnstein, R. J., & Loveland, D. H. (1964). Complex visual concept in the pigeon. *Science, 146,* 549–551.

Immelman, K. (1972). Sexual and other long-term aspects of imprinting in birds and other species. In D. S. Lehrmann, R. A. Hinde, & E. Shaw (Eds.), *Advances in the study of behavior* (Vol. 4, pp. 147–174). New York: Academic Press.

Kenward, R. E. (1978). Hawks and doves: Factors affecting success and selection in goshawk attacks on woodpigeons. *Journal of Animal Ecology, 47,* 449–460.

Lambert, J.-L. (1980). Stimulus fading procedures and discrimination learning by retarded children. In J. Hogg & P. S. Mittler (Eds.), *Advances in mental handicap research* (Vol. 1, pp. 83–198). New York: Wiley.

Lorenz, K. Z. (1935). Der Kumpan in der Umwelt des Vogels. *Journal für Ornithologie, 83,* 137–213.

Moltz, H., & Stettner, L. J. (1961). The influence of patterned-light deprivation on the critical period for imprinting. *Journal of Comparative and Physiological Psychology, 54,* 279–283.

Pietrewicz, A. T., & Kamil, A. C. (1981). Search images and the detection of cryptic prey: An operant approach. In: A. C. Kamil & T. D. Sargent (Eds.), *Mechanisms of foraging behavior.* New York: Garland.

Ramsay, A. O. (1951). Familial recognition in domestic birds. *Auk, 68,* 1–17.

Ryan, C. M. E. (1982a). Concept formation and individual recognition in the domestic chicken. *Behaviour Analysis Letters, 2,* 213–220.

Ryan, C. M. E. (1982b). *Mechanisms of individual recognition in birds.* Unpublished masters thesis, University of Exeter.

Salzen, E. A., & Meyer, C. C. (1968). Reversibility of imprinting. *Journal of Comparative and Physiological Psychology, 66,* 269–275.

Schein, M. W. (1963). On the irreversibility of imprinting. *Zeitschrift für Tierpsychologie, 20,* 462–467.

Terrace, H. S. (1963). Discrimination learning with and without errors. *Journal of the Experimental Analysis of Behavior, 6,* 1–27.

Terrace, H. S. (1966). Stimulus control. In W. K. Honig (Ed.), *Operant behavior: Areas of research and application.* New York: Appleton-Century-Crofts.

Terrace, H. S. (1972). By-products of discrimination learning. In G. Bower & J. Spence (Eds.), *The psychology of learning and motivation* (Vol. 5). New York: Academic Press.

Vaughan, W., & Greene, S. L. (1984). Pigeon visual memory capacity. *Journal of Experimental Psychology: Animal Behavior Processes, 10,* 256–271.

Vidal, J.-M. (1976). L'Empreinte chez les animaux. *La Recherche, 63,* 24–35.

6 Configural Processes in Pigeon Perception

Kenneth M. Steele
Duke University

Work on pattern recognition pushes us into the problem of the definition of a stimulus. One approach has been to assume that a stimulus is best considered as a collection of smaller units or elements commonly termed *features*. Experimental work centers around the discovery of such units, or their manipulation. A major purpose of this chapter is to question this approach. It is suggested that features do not provide a satisfactory unit of analysis because they lack autonomy where feature theory would expect such to occur. It is further suggested that the reason for this lack of autonomy is that features represent the products of other basic processes. The problem of the definition of a stimulus is better approached as one of differentiation: what controls the manner in which the stimulus is ''parsed'' or ''carved-up'' into units; whether these things are called features or objects. From this view features will be ''flexible''; they will appear and disappear as conditions change. Experimental analysis should focus on the isolation of the conditions that control differentiation. Thus, the specification of a stimulus becomes the description of the engagement of the processes of differentiation.

PERCEPTUAL FEATURES AS UNITS

From the widespread use of the term, a reader would assume that there was universal assent that a stimulus is best analyzed as being composed of some smaller number of elements termed features. But, exactly, what is a feature? The attempt to find a precise definition is likely to turn out to be like the never-ending task of completing a circle along a Mobius strip.

Features may be defined as units familiar from high school geometry (e.g., points, lines, and angles), or they may be given quasitopological designations (e.g., connectivity, open-right, or vertical symmetry: See Blough, 1982, 1985; Hyrcenko & Harwood, 1980; Morgan, Fitch, Holman, & Lea, 1976). The presence of certain units may be both necessary and sufficient, or identification may be more democratic ("polymorphous" sets: see Morgan et al., 1976).

The identification of what is meant by features is, itself, apparently polymorphous. One reason for the imprecision is that the term is used in two very different senses. The first use is a convenient and conventional label for talking about control by portions of a display. From this use, when it is said that two classes are discriminated on the basis of some feature, it is meant that an identifiable subportion of the display comes to control much of the performance in a particular task. Nothing is implied necessarily about the perceptual mechanisms that are the basis of this performance. Hence it would not be an oxymoron for even the most diehard Gestalt theorist to speak of features in this manner.

The second use of the term feature implies a theory about the mechanism of perception. In this case, an object is literally the collection of particulate subunits to which our perceptual apparatus is sensitive. Different objects are more or less discriminable depending on the number of elements they share. The continued impact of Neisser's (1967) text, *Cognitive Psychology,* and the neuroanatomical theorizing of Hubel and Wiesel (1965) has much to do with the widespread appeal of this view.

This imprecision has produced a problem. Because we are often able to talk about features (i.e., the first use) in our description of performance, there is a tendency to confound this with the latter position that this also describes the basis of perception. One example is Lea's (1984) analysis of discrimination performance by pigeons in Herrnstein-type visual concept experiments.

Two obvious issues arise for feature theory: the identification of the units, and the demonstration of their particulate nature. The issue of the particulate nature of stimuli for the perceptual apparatus is vitally important because it provides the justification for attempts to decompose stimuli, like Blough's (1985) recent attempts to derive a feature model of his pigeon's performance. Among current perceptual research with pigeons, the work by John Cerella is most explicit on this issue.

Cerella's (1982) argument is that stimuli, like the "summer vacation" photographic slides used by Herrnstein and colleagues, are collages of particulate fragments for the pigeon. Even the most apparently sophisticated and human-like discrimination by pigeons represents a solution based on static local cues. He cites two kinds of evidence. The first is the lack of difference in performance between discrimination tasks that do or do not contain additional structural information. For example, Cerella (1977) found no evidence that pigeons were sensitive to the representation of perspective in static wire-frame displays of cubes. Instead, discriminative performance was not much different from a similar task in which the lines were presented in regular but formally nonperspectival arrangements. Cerella (this volume) extends this analysis to other types of perspective information (e.g., rotation, reflection, dilation, and contraction).

The second kind of evidence is the lack of disruption of performance when the original training stimulus is changed, while preserving component parts. For example, Cerella (1980) trained pigeons to discriminate the cartoon character "Charlie Brown" from other members of the "Peanuts" menagerie. He found that performance was unaffected by the presentation of scrambled-Charlies; that is, to scrambled rearrangements of thirds of the original training stimulus. Thus the pigeons were insensitive to the arrangement of the entire original. Instead, performance depended on the presence of the component parts that were preserved within the thirds that were scrambled.

Feature theories, such as Cerella's, make an obvious prediction in certain types of transfer to new discrimination tasks. Successful training in the original task shows that the positive and negative stimulus classes differ in their composition for the bird. If those same features, and their significance, remain present in a new task, then there should be a ready transfer of successful performance to the new discrimination task. In other words, performance should show positive transfer. Therefore, in such cases, it is significant to find failure of positive transfer.

Failure to Obtain Positive Transfer

Four pigeons were trained to discriminate between individual occurrence of the upper case alphabetic letters, *D* versus *X,* using a multiple random-interval extinction food-reinforcement schedule. Hence, each session consisted of successive presentations of D or X, in which one letter signaled the opportunity to obtain food (S+), and the other signaled no food (S−).

The top row of Fig. 6.1 gives an approximate view of the original training displays. The rectangle indicates the border of the 8-by 5-centimeter (cm) response key. Each letter was approximately 5 by 7 millimeters (mm) in size, and appeared as white on dark gray. For 2 pigeons X served as S+, whereas D served as S+ for the other 2 birds. Each presentation of a letter was chosen to randomly occur in 1 out of 45 possible locations that formed a 9 by 5 matrix on the response key.

After discriminative performance was stable, the food significance of the letters was reversed, and the pigeons were trained under the new discrimination. This reversal procedure was continued until it took less than about five sessions for performance to reach asymptote after such a reversal. The top row of Fig. 6.2 shows performance by one of the subjects, Bird 159, over the later stages of this condition.

Performance per session is plotted in terms of a discrimination ratio, which represents the proportion of S+ responses out of total responses. The practical range of the measure here is from 0.5 (indifference) to 1.0 (only S+ responses).

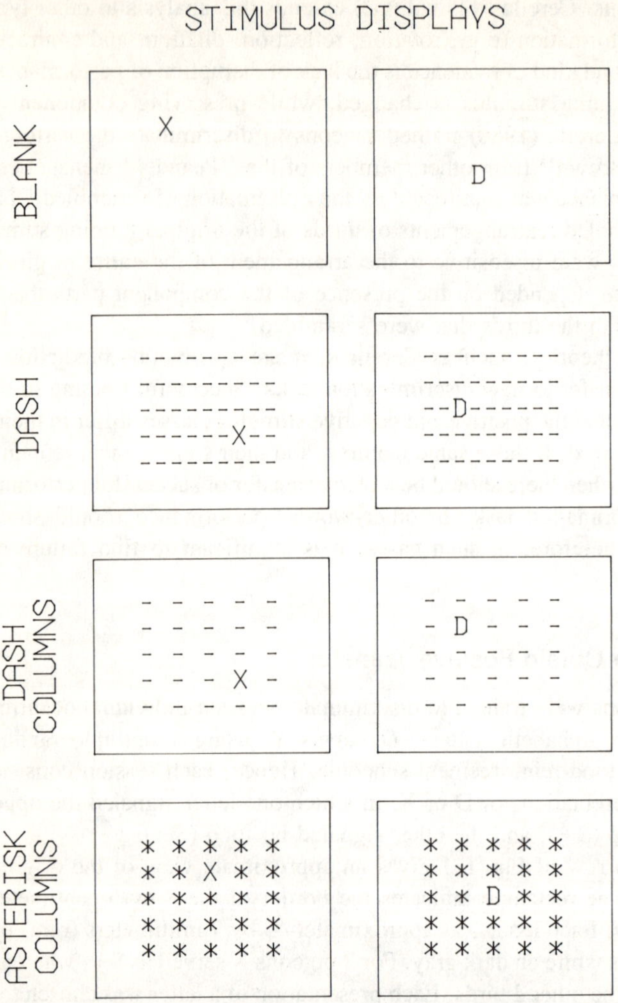

FIG. 6.1. Stimulus displays for Bird 159. Each row shows a pair of displays from a training condition. The name of the training condition is on the left side and refers to the additional stimuli that accompanied presentation of the discriminative stimulus (X or D). The rectangular border represents in outline the shape of the response key.

The top row begins with the return to the original S+, which was X for Bird 159. The top row then shows performance across conditions when X was S+ (Orig) and D was S+ (Rev). The rapid improvement in performance across successive reversals shows that Bird 159 could readily discriminate between the two letters.

In the next phase, the displays were changed so that the D or X now appeared

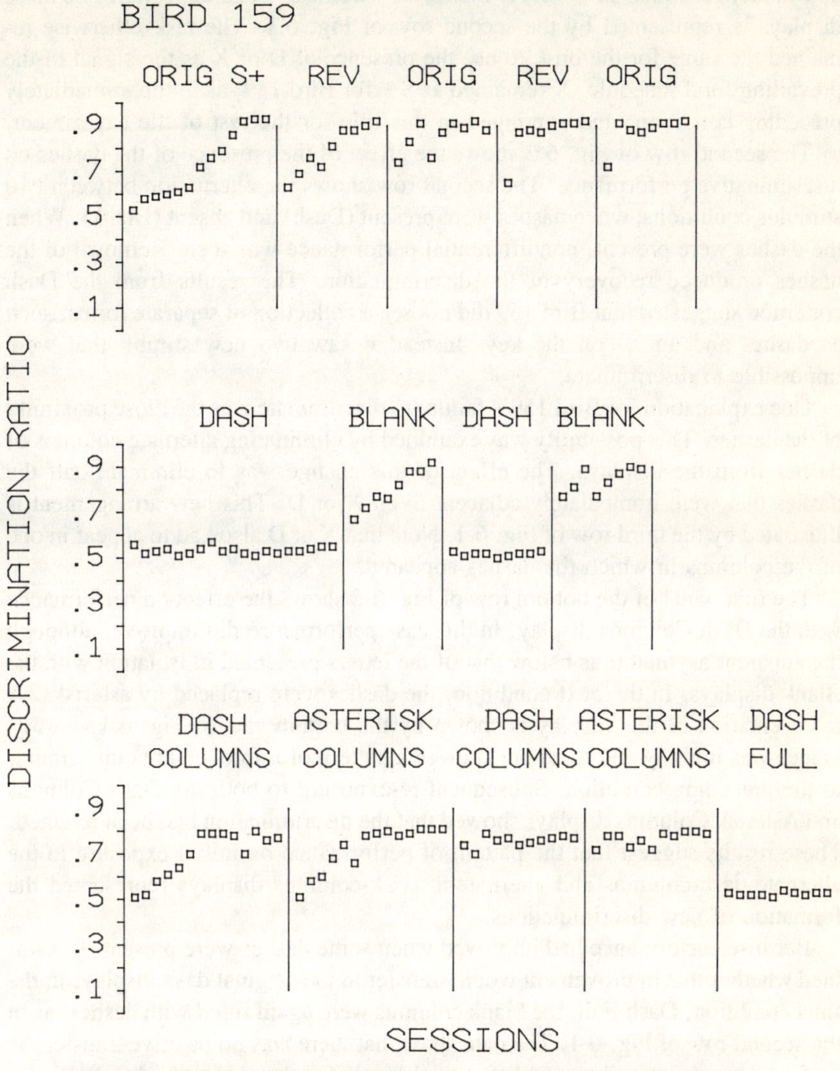

FIG. 6.2. Discrimination performance by Bird 159. Each point shows the discrimination ratio value for a single session. A discrimination ratio is the proportion of S+ responses out of total responses. Vertical lines indicate a change in conditions. The top row shows performance when X was S+ (Orig) and D was S+ (Rev). No other forms were presented on the response key for those conditions. The bottom two rows show performance when only X was S+, and the presence of additional stimuli on the response key was manipulated. The stimulus condition in effect is indicated just above each condition.

at a random location in a 9 by 5 matrix of "dashes." The appearance of these displays is represented by the second row of Fig. 6.1. The task otherwise remained the same for the bird: to use the presence of D or X as the signal of the prevailing food schedule. X remained as S+ for Bird 159, as in the immediately preceding condition, and continued in this role for the rest of the experiment.

The second row of Fig. 6.2 shows the effect of the presence of the dashes on discriminative performance. The second row shows the alternation between two stimulus conditions: when dashes were present (Dash) and absent (Blank). When the dashes were present, nondifferential performance was seen. Removal of the dashes produced recovery of the discrimination. The results from the Dash condition suggested that Bird 159 did not see a collection of separate forms, such as dashes and an X, on the key. Instead it saw two new stimuli that were impossible to discriminate.

One explanation for Bird 159's failure to discriminate was the close proximity of the dashes. This possibility was examined by eliminating alternate columns of dashes from the displays. The effect of this change was to eliminate half the dashes that were immediately adjacent to an X or D. This new arrangement is illustrated by the third row of Fig. 6.1. Note that X or D also had to appear in one of the columns in which the dashes appeared.

The first panel of the bottom row of Fig. 6.2 shows the effect on performance with the Dash Columns display. In this case performance did improve, although the apparent asymptote is below that of the letters presented in isolation with the Blank displays. In the next condition, the dashes were replaced by asterisks. A representation of this display is shown by the bottom row of Fig. 6.1. Performance was initially disrupted but showed a pattern of improvement quite similar to the preceding condition. Subsequent re-exposure to both the Dash Columns and Asterisk Columns displays showed that the discrimination had been retained. These results suggest that the pattern of performance on initial exposure to the alternate-dash-columns and alternate-asterisk-columns displays represented the formation of new discriminations.

Because performance had improved when some dashes were present, I examined whether this improvement would transfer to the original dash display. In the final condition, Dash Full, the blank columns were again filled with dashes, as in the second row of Fig. 6.1. The data show that there was no positive transfer, as performance decreased to nondifferential levels for the entire condition.

The significance of this pattern of results is the difficulty produced for theories that presume a stimulus is composed of autonomous subunits. However many features one would speculate constituted an X or a D, the success with the original discrimination training showed that the letters were readily differentiable. It may be presumed that Bird 159 would continue to use the difference(s) between letters, if possible, in the latter conditions. The bird, however, was unable to do so when the letters were surrounded completely by dashes.

It is not just the failure to discriminate when dashes were present, but the occurrence of this failure given there was discrimination when the dashes occurred in alternate columns. I had chosen to use dashes because they appeared very dissimilar to the X and D. If discrimination had failed under all conditions, then it would be appropriate to argue that my judgment of dissimilarity for the pigeons had failed. In other words, it would be parsimonious to claim that a dash and an X did contain common features so as to render them indiscriminable. But the loss of discriminative performance produced by the inclusion of additional dashes strongly suggested that the two discrimination tasks, Dash and Dash Columns, were different tasks, although comprised of the same constituents. Hence, performance was not predicted by knowledge of the constituents alone.

The justification for features as theoretical units has been based on their atomic status, i.e., that they provide the elements out of which perception could be built. It is their lack of autonomy under conditions such as the preceding that argues against that justification.

One alternative is to suggest that features represent products, not elements. In other words, we restrict the meaning of feature to that of a label of performance. Instead, the question becomes one of differentiation. What differences are necessary in the relationship between the properties of the distribution of lighted pixels that constitute the irrelevant ''ground'' (i.e., the dashes) and those that constitute the ''figure'' (i.e., the X) to produce their separation? The presence of a thing is the product of these relationships, whether we call these things features or objects. Hence, the primary data of interest become the description of the relationships that control differentiation.

From this viewpoint, the meaning of the preceding effects may be approached in the following manner. Given that X and D are no longer seen, it may be assumed that new things are seen. What relationships in the display contribute to the disappearance of the old stimuli or their absorption into new stimuli?

AN ANALYTIC TECHNIQUE: THE INTERFERENCE TASK

One inspiration for experimental analysis comes from what perception theory distinguishes as unitary, or *integral,* and *separable* stimuli (Garner, 1974; Lockhead, 1966). One may define an integral stimulus by its relative resistance to individuation or decomposition. If a stimulus array is integral, then it is difficult to ''pay attention'' to one aspect of a stimulus without attending to its other aspects, whereas an array that is separable allows independent attention to its constituents. The utility of this is seen in a *selective-attention* task, where the subject is asked to classify stimuli on the basis of a particular (relevant) dimension, which is accompanied by a second (irrelevant) dimension. The extent to which performance is a function of the manipulation of the *irrelevant* dimension is a measure of the unitary nature of the stimulus array. As applied to the performance of Bird 159, the X versus D discrimination was controlled by the manipulation of the irrelevant "dimension" of spacing/number of the dashes.

This distinction can be used to advantage in the analysis of form perception.[1] The manipulation of nominally irrelevant dimensions is used to discover those relations that control the availability of particular forms. To put this basic idea to test, I developed a selective-attention-type procedure for pigeons that is termed an *interference* task. In this task, the question is whether manipulation of aspects of irrelevant forms control an extant discrimination. When such manipulations are effective in controlling performance, the aspects are not perceptually irrelevant.

A subject is instructed first on the relevant task by a simple discrimination procedure. For example, pigeons could be trained to discriminate between the two letters, *R* versus *P*, by being given food for pecking in the presence of one but not the other. Following the development of discriminative performance between these *target* items, an additional *distractor*, such as the letter *B*, is introduced onto the display. The pigeon's task is to continue to attend to the original forms that signal the food schedule and respond appropriately. The effect of interest is whether the pattern of disruption of discriminative performance (i.e., interference) covaries with the systematic manipulation of characteristics of the distractor's presence.

Test of the Method

This experiment examined whether the factors of proximity and similarity of distractors to targets controlled the integrality of the stimulus for a pigeon. The influence of both factors in human perception is well known (Estes, 1972, 1974; Pomerantz, 1981). It is increasingly difficult to find or attend to one constituent of an array as other portions become both more similar and proximate. If comparable effects occur with pigeons, then, in this task, stronger interference with established performance will result from more similar and more proximate distractors.

Six pigeons served as subjects. All had prior experimental experience with similar kinds of tasks, but none with the specific stimuli used here. The birds were kept at 80% of ad libitum weight for the experiment.

The apparatus consisted of a typical operant-conditioning chamber, a video monitor, and a lens to focus the monitor's display onto the 8-by 5-cm response key. A 2-cm-deep hood around the response key prevented the adoption of extreme viewing angles. A microcomputer, located in a nearby room, controlled the session sequence, presentation of stimuli, and data recording.

---

[1]There are a series of effects that should be shown before one would technically term a stimulus integral. Here I use the spirit of the distinction. Cook (1983) has explicitly tested for integral stimulus effects in a matching-to-sample procedure with pigeons and reported positive results.

The stimuli consisted of the upper case letters *R, P* (targets) and *H, B, Z* (distractors). Each letter was taken from the microcomputer's character set. On the response key, each letter was white (on dark gray), inverted (due to projection), and occupied approximately a 5-by 7-mm area.

The procedure consisted of two phases: training and interference. The training phase consisted of establishing the discrimination between R and P under a multiple schedule. Each session consisted of presentations of R and P on the otherwise-blank response key. The letters were presented individually in a sequentially balanced pseudorandom order. The order was varied from session to session. Each presentation of R or P lasted 40 seconds. There were 72 presentations per session, and each presentation was separated by a brief 2-second interval during which the response key was dark. Each letter was presented equally often in one of the 6 possible locations, which formed a T-shaped matrix that was used in the interference phase. R was the S+ and signaled a random-interval 20-second food-reinforcement schedule. Reinforcement consisted of 3 seconds access to mixed grain. P was the S− and signaled that food was not available (extinction). This phase lasted 10 sessions. The interference phase held constant all conditions from the previous phase, with the exception that letters from the distractor set were also projected onto the display.

The influence of proximity was investigated by manipulating the interletter distance between a constant number of letters. The stimulus display for each of the components consisted of a single target letter and two occurrences of one of the distractor letters in a triangular arrangement. The target letter appeared equally often at each vertex. The interletter distances could assume one of three values during a session. These are illustrated in Fig. 6.3. They are, hereafter, referred to as "close," "medium," or "wide." Close spacing represented a separation of the "base" letters of the triangle by a single blank 5-by 7-mm character space, with the "apex" letter of the triangle immediately underneath that blank space. Medium spacing was produced by moving the position of the apex letter one character space vertically down from the centerpoint of the display. Wide spacing was produced by the additional horizontal expansion of each base letter one character space from the centerpoint. There was an equal number of instances of each type of spacing per session, and these were equally split between S+ and S− components. The presentation order of proximity differences over a session occurred in a sequentially balanced pseudorandom order, and this order was varied from session to session.

The influence of target–distractor similarity was investigated by the use of different letters as distractors. The criterion for a similarity effect was that the amount of disruption produced by the different distractor letters maintain their same rank order across the interference phase. The particular yardstick along which one arranges letters by similarity is not yet agreed on. Blough (1985) discusses some of these models. But the assumption of ordinality underlies all specific models of letter similarity and therefore provides the more general test.

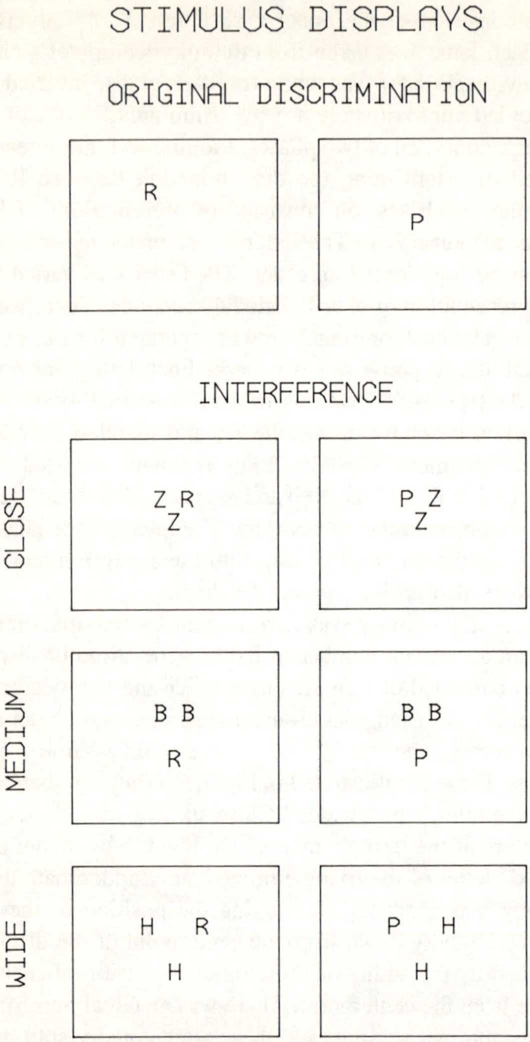

FIG. 6.3. Stimulus displays in the interference task. The top row represents stimuli from the training phase. The bottom three rows represent displays used in the interference phase. Close, medium, and wide show the spacing used for the proximity manipulation. Z, B, and H were the distractor letters used for the similarity manipulation. Only one type of distractor letter was used within a single session, and was presented under all spacing conditions. The entire set of distractors was presented in a repetitive sequence across successive sessions.

Only one particular letter was used as a distractor per session. Thus, three sessions (a "cycle") were needed to expose the subjects once to the entire set of distractors. The presentation order of distractors across sessions that comprised a cycle (e.g., H → B → Z vs. Z → B → H) was unique for each bird, but the same sequence-of-sessions per cycle was used for any given bird throughout the interference phase.

Discrimination ratios were computed for individual subjects for each session. The overall averaged discrimination ratio for the 6 birds for the last 5 sessions of the training phase was .84. The range was .72 to .89, with only one bird below .80. These values show that the birds discriminated satisfactorily between R and P.

Figure 6.4 shows the proportional change in performance from the training phase for the entire interference phase. The left panel shows the amount of interference as a function of which letter served as distractor. The right panel shows the amount of interference as a function of interletter spacing. Both panels are plotted for "change ratio" across cycles (7 cycles or 21 sessions total). The

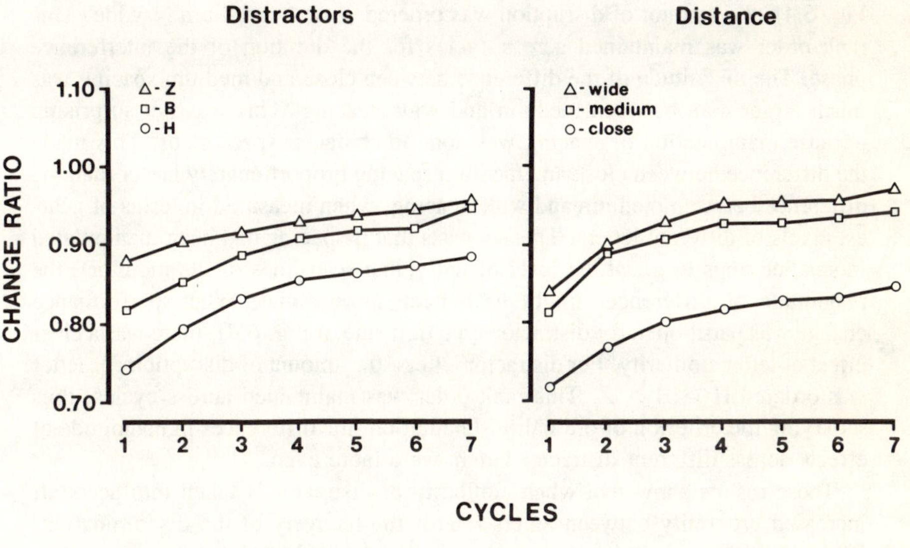

FIG. 6.4. Averaged performance in the interference phase. Change-ratio values show the normalized proportional disruption from performance in the original single-stimulus training phase. Proportional change in performance was determined for individual birds and then averaged across birds. Unaffected performance would produce a value of 1.00 and loss of discrimination would produce values less than 1.00. A cycle is the three-session block that comprised exposure to the entire set of distractor letters. The left panel shows performance as a function of the different distractor letters (the similarity variable). The right panel shows performance as a function of interletter spacing (the proximity variable).

change ratio was computed by dividing a bird's discrimination ratios for an interference-phase session by its corresponding averaged discrimination ratios from the training phase (last 5 sessions). Therefore, change ratios represent the change in performance for each individual bird based on its prior performance. These normalized data then were averaged over the 6 birds to produce a summary figure. These averaged data were representative of the individual performances by all pigeons. A change ratio value of 1.00 indicates no disruption of performance, and values of less than 1.00 indicate a decrease from baseline-training performance.

Both sets of functions show maximum disruption occurred at the beginning of the interference phase. At this point, absolute response rates for S+ and S− showed a convergence. Response rates in S+ declined but increased in S−. The recovery pattern suggests that the discrimination must be reformed following the introduction of the distractors. These changes are best seen as "learning curves."

Both proximity and similarity of the distractor letters affected the pigeon's ability to respond on the basis of the target alone. For proximity (right side of Fig. 6.4), the amount of disruption was ordered close > medium > wide. This rank order was maintained across cycles for the duration of the interference phase. The magnitude of the difference between close and medium spacing was much larger than between medium and wide spacing. This was not surprising because manipulation of spacing was done in character-space steps. This made the difference between close and medium spacing proportionately larger than the difference between medium and wide spacing, when measured in terms of nearest pixels of different letters. This suggests that if spacing had been manipulated in smaller steps (e.g., at the level of individual scan-lines of the monitor), the magnitude of differences might have been more equal. When performance change was partitioned for distractor-type (left side of Fig. 6.4), there was a clear effect of letter similarity. For distractor letters, the amount of disruption per letter was ordered H > B > Z. This rank order was maintained across cycles, thus satisfying the criterion of ordinality. In addition, the differences in magnitude of effects across different distractor letters were more even.

These results show that when similarity of distractor is taken into account, increased proximity between letters retards the recovery of the discrimination. When interitem proximity is taken into account, similarity of distractor retards the recovery of the discrimination.

The effect of proximity supports the interpretation applied to the performance of Bird 159. It is important to remember that the number of items was held constant here, and *only* spatial relationship was varied. This result argues strongly against accounts of discrimination that are based on autonomous features. Equally of interest was that a clear effect was produced by the addition of only two distractors into a display. This indicates the ease with which such effects were produced.

It could be suggested that the different spatial conditions created different sets of features. In other words, there were *emergent* features that were responsible for the different performance under the different spacing conditions. This is, however, just a weaker version of the argument suggested here. The emergent feature argument states that some combinations of features produce new types of features. For example, there is some perceptual benefit of triangles that is not predicted from the knowledge of performance with only lines of the appropriate orientation (Treisman & Gelade, 1980; Treisman & Paterson, 1984). The emergent feature argument acknowledges implicitly the point specified earlier: Relationships in the display determine what is available. This position should specify the same course of research.

The finding of an effect of letter similarity on performance replicates, with a free-operant procedure, Blough's (1982) more extensive discrete-trial analysis of similarity relations of letters for pigeons. Of the distractor letters, H produced the strongest disruption here, whereas Blough's cluster-analysis data would suggest that B should have produced the stronger interference effect. The reason for the difference is unclear. It may be due to small but important differences in the character sets used by his computer and mine.

The proximity and similarity variables were treated as if they were orthogonal, e.g., like wavelength and intensity of light. Is this assumption true? Probably not. The similarity variable provided a means to deal with the nonarbitrary set of forms stored in the character chip in my microcomputer. These forms have been shaped by people for specific reasons. One reason, presumably, is to make such forms separate and distinct. At bottom, the R on the computer screen is a collection of dots. The variables of pixel distribution that make R an entity are likely of the same class as made different arrangements in the preceding three-form experiment more or less discriminable. This implies that both the proximity and similarity effects here depend on a common set of variables. This suggests that the next step is to focus directly on the discriminability of various rules of pixel distributions. In other words, progress is likely to be made by the investigation of textures of the kind pioneered by Julesz (1971, 1981). In addition, it may be of great help that such stimuli are mathematically specified and therefore more precisely reproduced in different laboratories.

If it is true that the same class of variables underlies the proximity and similarity effects, what does this interpretation say about the meaning of the interference procedure. It suggests that it may be incorrect to consider the task as just an "embedded-figure" problem. In other words, the pigeon is not learning to "find the hidden R." Instead, the procedure is one of stimulus formation; it creates several new stimuli that the pigeon learns to discriminate. Hence, the Z-R-Z versus Z-P-Z discrimination was separate to some extent from the H-R-H versus H-P-H discrimination; and, in addition, the latter was also more difficult. This speculation is supported by the effects noted with Bird 159. Remember that the bird was able to solve the discrimination when the dashes were presented in alternate columns, but not when all columns were again filled with dashes. This suggests that the two discrimination tasks were different, although comprised of the same constituents.

The last two paragraphs have been broad speculation. They serve the purpose of suggesting a fruitful path for research. The alternative is to produce more and more specific lists of features in a particular discrimination task. The consequence of that approach is the strong likelihood that the list will have to be radically changed when a new discrimination task is examined (cf. Herrnstein's, 1984, discussion of "flexible features"). Features exist, but let us keep them in their proper role.

ACKNOWLEDGMENTS

This research was supported by National Research Service Award F32 MH 09235 from NIMH, and by NSF grants to Duke University. I thank R. J. Herrnstein for helpful comments on an early version of this chapter.

REFERENCES

Blough, D. S. (1982). Pigeon perception of letters of the alphabet. *Science, 212,* 397–398.

Blough, D. S. (1985). Discrimination of letters and random dot patterns by pigeons and humans. *Journal of Experimental Psychology: Animal Behavior Processes, 11,* 261–280.

Cerella, J. (1977). Absence of perspective processing in the pigeon. *Pattern Recognition, 9,* 65–68.

Cerella, J. (1980). The pigeons analysis of pictures. *Pattern Recognition, 12,* 1–6.

Cerella, J. (1982). Mechanisms of concept formation in the pigeon. In D. J. Ingle, M. A. Goodale, & R. J. W. Mansfield (Eds.), *Analysis of visual behavior* (pp. 241–260). Cambridge, MA: MIT Press.

Cook, R. G. (1983). *The effects of compound stimulus structure on dimensional discriminations by pigeons and humans.* Unpublished doctoral dissertation, University of California, Berkeley.

Estes, W. K. (1972). Interactions of signal and background variables in visual processing. *Perception & Psychophysics, 12,* 278–286.

Estes, W. K. (1974). Redundancy of noise elements and signals in the visual detection of letters. *Perception & Psychophysics, 16,* 53–60.

Garner, W. R. (1974). *The processing of information and structure.* Potomac, MD: Lawrence Erlbaum Associates.

Herrnstein, R. J. (1984). Objects, categories, and discriminative stimuli. In H. L. Roitblat, T. G. Bever, & H. S. Terrace (Eds.), *Animal cognition* (pp. 233–262). Hillsdale, NJ: Lawrence Erlbaum Associates.

Hubel, D. H., & Wiesel, T. N. (1965). Receptive fields and functional architecture in two nonstriate visual areas (18 and 19) of the cat. *Journal of Neurophysiology, 28,* 229–289.

Hyrcenko, O., & Harwood, D. W. (1980). Judgments of shape similarity in the Barbary dove (*Streptopelia risoria*). *Animal Behavior, 28,* 586–592.

Julesz, B. (1971). *Foundations of cyclopean perception.* Chicago: University of Chicago Press.

Julesz, B. (1981). Figure and ground perception in briefly presented isodipole textures. In M. Kubovy & J. R. Pomerantz (Eds.), *Perceptual organization* (pp. 27–54). Hillsdale, NJ: Lawrence Erlbaum Associates.

Lea, S. E. G. (1984). In what sense do pigeons learn concepts? In H. L. Roitblat, T. G. Bever, & H. S. Terrace (Eds.), *Animal cognition* (pp. 263–276). Hillsdale, NJ: Lawrence Erlbaum Associates.

Lockhead, G. R. (1966). Effects of dimensional redundancy on visual discrimination. *Journal of Experimental Psychology, 72,* 95–104.

Morgan, M., Fitch, M., Holman, J., & Lea, S. E. G. (1976). Pigeons learn the concept of an "A." *Perception, 5,* 57–66.

Neisser, U. (1967). *Cognitive psychology.* New York: Appleton-Century-Crofts.

Pomerantz, J. R. (1981). Perceptual organization in information processing. In M. Kubovy & J. R. Pomerantz (Eds.), *Perceptual organization* (pp. 141–180). Hillsdale, NJ: Lawrence Erlbaum Associates.

Treisman, A., & Gelade, G. (1980). A feature-integration theory of attention. *Cognitive Psychology, 12,* 97–136.

Treisman, A., & Paterson, R. (1984). Emergent features, attention, and object perception. *Journal of Experimental Psychology: Human Perception and Performance, 10,* 12–31.

II SHAPE AND FORM

7 Form Similarity and Categorization in Pigeon Visual Search

Donald S. Blough
Brown University

The first question of form recognition is psychophysical: How are objective measures of stimulus objects related to subject classifications of these objects? For empirical testing, this question is often posed in terms of similarity between pairs of forms, which may conveniently be measured in many ways. The method of measurement used here is a bit uncommon for the purpose. Pairs of forms are compared by asking how rapidly one of the two can be found when it is surrounded by many copies of the other. If search is slow, the forms are, by definition, similar; if rapid, they are dissimilar.

Posing questions of form perception in terms of similarity, pair comparison, and search naturally highlights certain questions and perhaps obscures others. For example, similarity suggests symmetry: It may seem that object A should be as similar to object B as B is to A. Measurements do not always confirm this intuition, however (Tversky, 1977), and two interesting failures to confirm symmetry are noted later. Again, visual search may seem an exotic way to study similarity; how do its results compare to those from more conventional methods? This question also is briefly considered later. Finally, the continuity of similarity measures and their bias toward symmetry might be expected to obscure categorical aspects of form perception, but similarity data can shed light on categorization, as exemplified below.

Despite its idiosyncrasies, the search method has proven to be an effective procedure. Pigeons learn search tasks easily and perform them with great accuracy, aided perhaps by their species-typical foraging predispositions. In most of our past work (D. Blough, 1977, 1979; P. Blough, 1984; Blough & Franklin, 1985), birds learned to peck at one or a few specific targets (such as the letter "A") and to avoid others. Recently, though, we have found that pigeons will perform a particularly useful task that does not require such differential learning (D. Blough, 1986). In this task, which we have called odd-item search, a large number of identical forms appear on a display screen. Among these the pigeon must find a single different form, the "odd-item." From trial to trial both forms vary; the target form on some trials becomes the background form on others. Rapid learning occurs despite the apparent resemblance of this task to oddity learning, which at best is very difficult for pigeons (for a discussion of this, see, e.g., D'Amato, Salmon, & Columbo, 1985).

Odd-item search is an efficient way to determine interpair similarities among a large set of forms. In the experiments described here, accuracy typically exceeds 95%, whereas search speed varies consistently as a function of the forms paired on any given trial. Instead of learning specific form discriminations one after the other, pigeons become equally familiar with all the forms. The method thus overcomes the effect of learning order, which may seriously confound the results of other procedures.

The experimental situation, complete with an example of odd-item search, appears in Fig. 7.1. Here a pigeon is finding an "H" among "+" forms, displayed on a small video monitor that is hooked to an ATARI personal computer. Opposed rows of photocells and infrared emitters (Clauson, Izatt, & Shimp, 1985) detect the pigeon's peck as having occurred in one of six vertical stripes across the screen.[1] A display of 32 forms is usually used, arranged in 4 by 8 matrix, with targets permitted only in the interior 2 by 6 positions. In a typical session, a pigeon sees all possible pairs from a group of 16 forms and is rewarded with food on a small percentage of trials (e.g., 10%) for pecking the odd form. The pigeon has 10 seconds to find this target, and incorrect pecks during this time are not counted. In theory, the bird could always peck across the screen until it came to the target, but this is not typical, because search times for very hard discriminations often exceed the 10-second limit. When search fails for 10 seconds, a default time of 10 seconds is recorded and the display is repeated. After two or three repeated failures, the target form appears on a highly discriminable background. Most other details of the procedure may be found in published work, especially D. Blough (1986).

A simple transformation converts search time to a convenient measure of similarity. Like other reaction times, search time is presumably composed of a relatively fixed duration that arises from motor processes and other rather uniform activities, and a variable quantity that reflects discriminative processing (D.

---

[1]This means that the pigeon could make a spurious correct response by pecking above or below the target form. Such responses were probably rare, given high overall accuracy. It may be inadvisable to include vertical position in the definition of a correct response, because previous research has shown vertical position of pecks in this situation to be variable and somewhat uncertain, partly due to the typical open-beak peck topography. It is prudent to keep targets near the horizontal meridian, because reaction time is relatively constant across this meridian but increases when the bird reaches up or down (D. Blough, 1979).

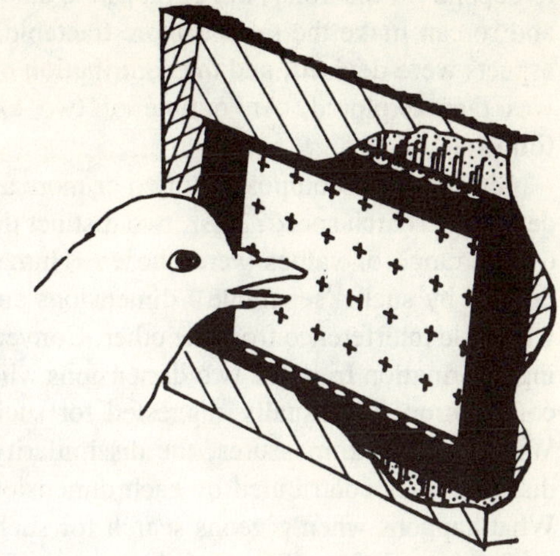

FIG. 7.1. Apparatus for search tasks, showing display monitor and photoelectric response-sensing system. A typical odd-item display appears here.

Blough, 1988). Because only the variable component is of interest here, a constant that estimates minimum search time is subtracted. For reasons unnecessary to recount, it also is helpful to use logarithms, and to subtract the measure from 1, the log of the 10-second maximum time allowed. One thus arrives at a measure of dissimilarity, D (equation 1):

$$D_{AB} = 1 - \log (RT_{AB} - K) \quad (1)$$

Here, A is the target form and B the background form, RT is geometric mean reaction time, and K is the estimated asymptote for the data set.[2] For each pair of forms A and B there are two values of D, one for A as target and B as background, one for B as target and A as background. Usually these values are comparable and therefore the measured similarity appears symmetrical. As mentioned before, symmetry sometimes fails in interesting ways.

COMBINATION RULES FOR FORM SIMILARITY

The search method has been applied to several questions relating to the psychophysics of form. Two of these questions assume an analytical approach: (1) of what units are forms composed, and (2) how do these units combine to determine the perceived similarity of forms? The answer to each of these questions appears

---

[2]This transformation has no effect on ordinal multidimensional scaling results and very little on clustering, but it seriously influences some explicit tests of combination rules (cf. D. Blough, 1972). Transformations akin to equation 1 have considerable use and justification in the reaction-time literature. Equation 1 is further supported by subsequent analyses; most notably, D is an almost linear function of the distances produced by the multidimensional scaling program ALSCAL from the data reported here (D. Blough, 1988).

to depend on the other, but perhaps the answer to the first can be predetermined and so can make the second more tractable. For this, stimuli with very distinct aspects were devised, and the contribution of these aspects to stimulus similarity was then explored. An outline of two experiments that take this approach follows.

Given stimuli composed of two or more aspects, how do these aspects jointly determine search speed? First, two distinct dimensions that varied independently over a range of values were chosen. Human subjects confronted with stimuli defined by such "separable" dimensions are able to attend to just one of them with little interference from the other. Conversely, subjects have trouble integrating information from the two dimensions when this might help in their task. The combination rule usually suggested for such stimuli is the "city-block" rule: With appropriate measures, the dissimilarity of two stimuli equals the sum of dissimilarities contributed by each dimension (e.g., Garner & Felfoldy, 1970). What happens when pigeons search for such stimuli?

The stimuli for experiment 1 appear in Fig. 7.2. Here, two variable aspects are literally separate, the line above varying in length and the black block below varying vertically in size. Each dimension assumes 4 values, yielding the 16 combinations shown. These can pair in 256 (16 × 16) ways, or in 240 when pairs of identical forms are omitted. Displays based on just 2 of these many pairs appear in Fig. 7.3.

After a few days of pretraining, 3 pigeons were run for 12 sessions. In each session each pair of forms appeared 12 times, and each of the 2 forms was the target on 6 of these.

It was first asked whether the specified dimensions systematically affect search speed. A convenient way to explore this question is through multidimensional scaling (MDS), which represents dissimilarity between stimulus objects as distance in a space of N dimensions (see Mumford, this volume; Shepard, 1980). A number of computer algorithms do this kind of scaling; the output shown in Fig. 7.4 came from ALSCAL (SAS Institute, 1982), which was programmed to assume only the monotonic rise of search speed with dissimilarity. Note that the similarity structure reflects the input variables. Each of two orthogonal dimen-

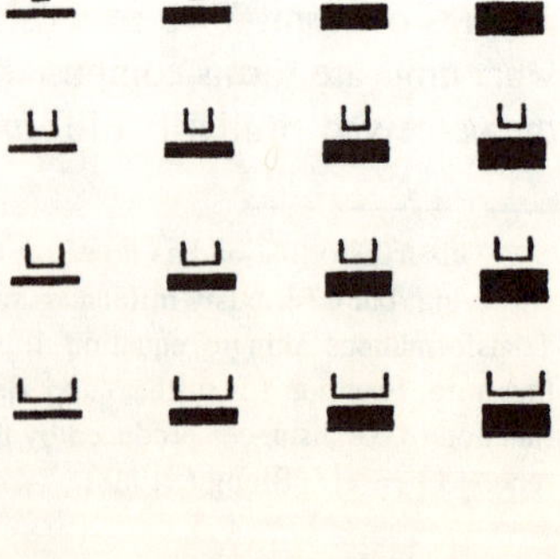

Fig. 7.2. The set of forms used in experiment 1. Two dimensions vary independently: The width of the U above, and the thickness of the black block below.

FIG. 7.3. Two sample displays from experiment 1. The pigeon was rewarded for pecking at the single different form, or "odd item." All possible pairs appeared in each session.

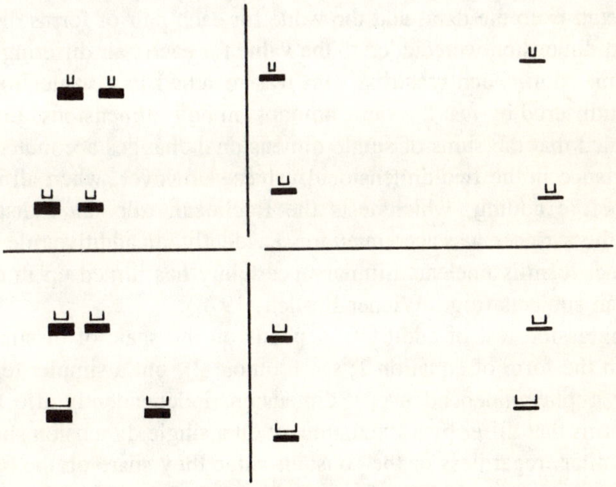

FIG. 7.4. Two-dimensional configuration of forms produced by ALSCAL from the search-speed data of experiment 1. The relative distance between pairs of forms in this diagram agrees well with the relative speed with which one of them was found on a background of the other.

sions is associated with one of the form variables, and forms that differ by several steps on one or two dimensions are farther apart than those that differ less. The fit was good, the derived distances accounting for 96% of the variance in optimally scaled data.

It is evident that the physical input dimensions are carried through the discriminative processes of search with sufficient independence to be recaptured from search speeds. Can anything further be said about these processes? By what rules do the dimensions jointly determine search speeds? This matter is too complex for a full review here, but let us consider several salient issues (cf. Beals, Krantz, & Tversky, 1968; Tversky, 1977; Wiener-Ehrlich, 1978). If dimensions were processed independently, the perceptual difference between stimuli that differ on two dimensions might be the simple sum of the differences on each dimension alone, as implied by the city-block rule mentioned earlier. Alternatively, if differences interfered somewhat with each other, dissimilarity might fall somewhat short of additivity. The ordinary Euclidean rule for distance could model such a case, because the hypotenuse of a triangle is shorter than the sum of its sides. Because Fig. 7.4 is drawn on a flat surface, the distances between forms conform to the Euclidean rule, and this could mean that similarity also conforms to this rule in the present case. Unfortunately, this strong conclusion cannot be justified by the scaling results alone. For example, city-block data can generate remarkably good Euclidean fits.

The additive model was tested more explicitly through linear regression. First, dissimilarities among pairs of forms that differed on just one dimension were selected from the data, and the value for each pair of forms differing only on the first dimension were added to the value for each pair differing only on the second dimension. Each resulting sum was matched to a value from a pair of forms that differed by just the same amounts in both dimensions. Linear regression revealed that the sums of single-dimensional changes accounted for 91.1% of the variance in the two-dimensional values. However, when all values were squared before adding, which tests the Euclidean rule, an indistinguishable 91.8% of the variance was accounted for. Evidently, an additive rule works well, but its exact form is unclear. Similar uncertainty has turned up in experiments with human subjects (e.g., Wiener-Ehrlich, 1978).

This regression test of additivity depends on the scale of dissimilarity used (that is, on the form of equation 1; see footnote 2), but a simpler test does not. Additivity implies independence; if dimensions independently affect similarity, pairs of forms that differ by a fixed amount on a single dimension should appear equally similar, regardless of the constant value they share on the other dimension. For example, consider the forms in Fig. 7.2. For any two rows, the forms directly above or below each other differ only in the width of the "U." If this difference is unaffected by the size of the black block, dissimilarities between these pairs of forms in two rows should all be the same. But they were not.

Consistently, when the constant aspect was at its largest or smallest value, search was a bit slower than for middle values. This interaction shows that simple additivity cannot entirely account for dimensional combination in this case.

Combination of Binary Features

The results of the first experiment suggest that additive differences can account for much, though not all, of the variance in the speed of search for stimuli defined by spatially distinct aspects. In this respect the results correspond with previous findings on human judgments of separable stimuli (e.g., Garner, 1978; Wiener-Ehrlich, 1978), and they suggest that this type of experiment may provide useful material with which to model the processing of highly analyzable stimuli. Does the additive rule apply when there are more stimulus aspects and these are not spatially separated? The next experiment examined this question.

In experiment 2, pigeons again searched for the odd item, with form pairs drawn from a set of 16, but here each stimulus assumed 1 of 2 values on each of 4 attributes. The set of forms is reproduced in Fig. 7.5. The attributes are color (actually, "outlined" or "solid"), size, shape, and orientation; all possible combinations are represented. The procedure was the same as in experiment 1, with search times recorded and the resulting 240 geometric mean reaction times reduced to a matrix of pairwise dissimilarity measures.

Multidimensional scaling again might seem an appropriate way to reduce these data, but four-dimensional plots are difficult to visualize, and for these data they were not very informative. In contrast, grouping through cluster analysis accounted for much of the variance and was easy to interpret. Figure 7.6 shows the results of hierarchical cluster analysis on these data (CLUSTER, SAS Institute, 1982). This program uses a distance algorithm to group forms that yield similar patterns of dissimilarities when paired with other forms. Clusters near the bottom of the diagram include quite similar forms; those near the top are increasingly dissimilar.

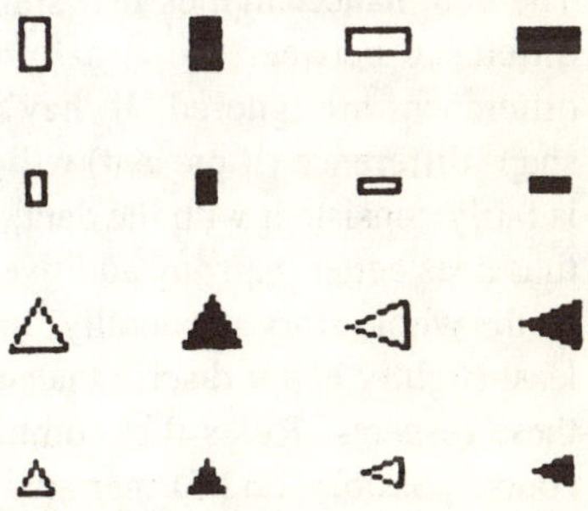

FIG. 7.5. The set of forms used in experiment 2. Two values of each of four attributes are combined in all possible ways: color, shape, size, and orientation.

The clusters rather neatly reflect the four specified attributes of the stimulus forms. As might be expected, the attributes vary in their influence on similarity. Color is most salient, for all clusters consist solely of black or of white forms. Shape is next most salient, for rectangles accompany triangles only at a high level. Then there is an interaction: Rectangles of different size are most similar, whereas triangles of different orientation are most similar. Also, groups of white forms are higher than comparable groups of black forms. This means that white forms are generally more distinguishable than black. For example, pigeons could find a white triangle among white rectangles faster than a black triangle among black rectangles. These indications from the cluster diagram were confirmed by specific quantitative comparisons of the single attributes. For example, where there was a single difference between forms, a target differing only in color from its background was more quickly detected than one differing only in form, and so on.

The additive model was fairly successful with the separable dimensions of experiment 1; how does it fare with the multiple attributes of experiment 2? As before, this can be tested by determining dissimilarities between forms that differ in a single way and adding these to predict dissimilarities between forms that differ in more than one way. It can already be seen that the additive prediction sometimes fails, because it is inconsistent with an interaction of attributes. Size and orientation differences cannot each add a constant to dissimilarity, because size differences are relatively more important for triangles and orientation differences are relatively more important for rectangles. This failure is confirmed by a quantitative test; simple addition of changes from single attributes accounted for only 53% of the variance in the dissimilarity of pairs differing in more than one way. This value, however, rose to 69% for the Euclidean distance rule, where dissimilarities are squared before combining. Though this is still not a very good fit, the improvement is in line with the notion that the combination of nonindependent or "integral" dimensions is best represented by the Euclidean metric.

The additive rule is but one of many possibilities, and Fig. 7.6 suggests an even simpler one, sometimes called the dominance rule (cf. D. Blough, 1972). The dominance implies that similarity is determined by only the most obvious difference between the two forms. If they differ in color, for example, other differences are ignored. If they are the same color, then the triangle–rectangle shape difference (if present) will determine speed, and so on. This simple model is fairly consistent with the data; it accounts for 81% of the relevant variance and thus does better than any additive variation. Nevertheless, dominance still cannot be the whole story: Generally, forms that differ in more than one respect were at least slightly better discriminated than forms differing in just the most salient of these respects. Rules that combine aspects of additivity and dominance are of course possible, and further analysis of these and other data may turn up appropriate modifications.

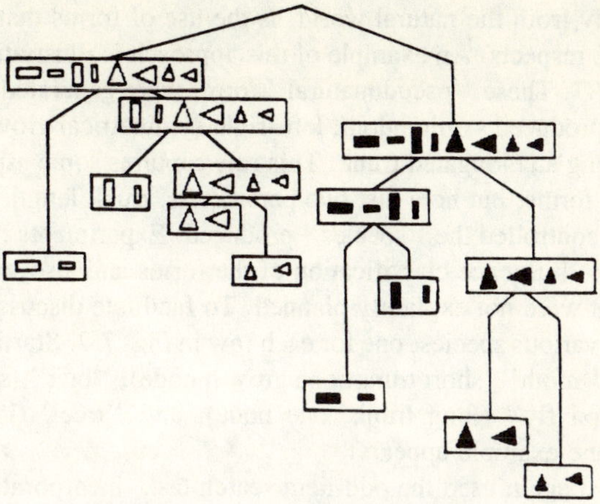

FIG. 7.6. Hierarchical clusters produced from search speeds in experiment 2. The height of a cluster is proportional to the logarithm of the average distance between its members.

Such combination rules may seem distressingly simplistic, given the sophisticated approaches to form perception outlined elsewhere in this volume, not to mention longstanding evidence for configural factors in form perception (e.g., Kubovy & Pomerantz, 1981). Before abandoning the combinatorial approach, however, one should recognize that predictive failure does not preclude the hypothesized combination processes. Failure may also indicate inappropriate scaling or inappropriate definition of the variables being combined. In other words, if additivity fails, the fault may lie not with additivity but with the units being added. Experimenter designation of attributes, features, or dimensions may not correspond to functional units. It is widely recognized that stimulus parsing continues to be an elusive problem.

Categories of Pseudonatural Objects

Other chapters in this volume discuss the categorization of natural objects. The problem of stimulus parsing is particularly acute for such objects. One can hardly imagine how such stimuli as "trees" or "fish" might be analyzed, or whether analysis is appropriate. Nevertheless, interpair similarities among many such stimuli might provide clues to answering these difficult questions, and interesting progress along these lines has been made using pictures (e.g., Sands, Lincoln, & Wright, 1982). Yet, the use of pictures poses substantial difficulties: It is hard to collect systematic variations that exclude irrelevant material, and because of the great complexity of most natural stimuli it is hard to specify them in objective terms.

A compromise designed to handle these difficulties, though obviously departing considerably from the natural world, is the use of forms that mimic natural objects in some respects. An example of this approach is illustrated by the set of forms in Fig. 7.7. These "pseudonatural" forms were generated by a computer algorithm that produced semirandom, left–right symmetrical growth from one or more nodes along an elongated trunk. This rule captures some aspects of growth in many living forms, but here just two parameters, trunk length and number of growth nodes, controlled the "species" produced. Experiments may reveal how these parameters influence classification of the forms and also whether subjects use aspects that were not explicitly planned. To facilitate discussion, names are attached to the various species, one for each row in Fig. 7.7. Starting with the top row, these are "moth" (short trunk, one growth node), "bug" (short trunk, two nodes), "dragon fly" (long trunk, one node), and "tree" (long trunk, two nodes—only one example appears).

Experiment 3 again used the odd-item search task, incorporating all possible pairs of the forms shown in Fig. 7.7. Of the analyses of dissimilarity scores, hierarchical clustering was most informative. These results are shown in Fig. 7.8. Notice that the pigeons only partially agreed with our species categorizations. All the bugs were classed together at an intermediate level (left), but body length seemed to make little difference. Instead, moths and dragon flies with smooth, simple profiles went together (lower right), whereas lacy wings characterized the moth and dragon fly in the center group. These were joined by the remaining dragon fly and the tree only at a high level, which implies that the latter were quite distinct from the other forms in the group.

This is a simple but promising beginning. Intended dimensions of variation are only partially reflected in the search data, and controlled randomness in the generation of the forms allowed the pigeons to tell us about an important feature that was not anticipated—"smooth" versus "lacy." Real, Iannazzi, Kamil, and Heinrich (1984) have reported that blue jays respond to a very similar difference in leaf damage ("neat" versus "messy") associated with different caterpillar prey. Surveys might turn up a relatively limited vocabulary of the most significant aspects within a given universe of forms.

Not evident in the clusters is a result that bears on the fundamental question of

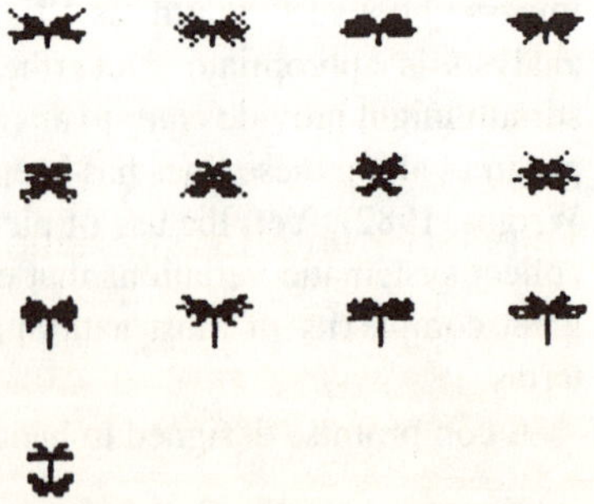

FIG. 7.7. The set of pseudonatural forms used in experiment 3. A different species appears in each row.

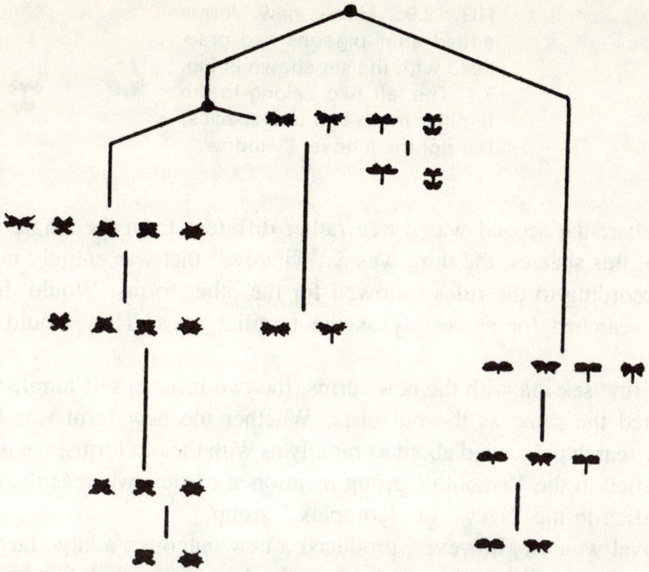

FIG. 7.8. Hierarchical clusters produced from search speeds in experiment 3. Clusters only partially correspond to species designations; "lacy wings" was a significant unanticipated feature.

symmetry. As already mentioned, similarity is often regarded as symmetrical: A is as similar to B as B is to A. The data for experiments 1 and 2 were quite symmetrical, with correlations of .96 and .95 between corresponding target–background and background–target pairs. Experiment 3 produced a notable exception to symmetry: Search was rapid when dragon flies (long tails) were paired as targets with other species but was slower when dragon flies were in the background. Here, the "tail" can be seen as a feature that is either present or absent, which suggests that it is easier to find a feature than it is to find the absence of a feature. This agrees with additional unpublished data from our lab, as well as similar results in human search (e.g., Julesz & Schumer, 1981; Treisman & Souther, 1985). The observation may be related to the well-known feature-positive effect in discrimination learning (cf. Hearst & Jenkins, 1974). Asymmetry in search might be a useful tool with which to explore the meaning and function of the elusive "feature."

Familiarity and Categorization

The preceding experiment with pseudonatural forms was the basis for an additional look at categorization. After the data shown in Fig. 7.8 had been collected, the pigeons continued their task with the same set of forms, except that three new forms were added (Fig. 7.9). One of these was a new moth, a type that should be

FIG. 7.9. Three new forms added after pigeons had practiced with the set shown in Fig. 7.7. The left two belong to the familiar moth and tree species; the right is a novel "window."

quite familiar; the second was a tree rather different from the single previous instance of this species; the third was a "window" that was entirely novel, not formed according to the rules followed for the other forms. Would these new forms be searched for as readily as the familiar ones? How would they be classified?

On the first session with the new forms, the two instances of familiar species were treated the same as the old ones. Whether the new form was target or distractor, search proceeded about as rapidly as with the old forms; the new moth was classified in the "smooth" group mentioned earlier, whereas the new tree was classified in the "lacy" or "complex" group.

The novel window, however, produced a new outcome: a large target-background asymmetry. When the window was background, search for the target (a familiar category) was considerably more rapid than for any other form combination. In contrast, when the window was the target, search averaged almost twice as long as for any other combination. Four days later search was relatively speedy with the window as either background or target, and cluster analysis placed it in a group of its own.

Evidently, form familiarity plays a role in odd-item search, though in principle this task does not require form memory. Furthermore, search for the pseudonatural forms appears to be categorical in the following sense. It is known that the new forms were discriminable from the old ones, because searches involving these forms were successful. Though discriminable from the old forms, the two new forms of familiar species were treated as old, whereas the form that was very different was at first rejected as a target. These results suggest that practice with the original stimuli defined a schema, or schemata, for targets. The asymmetry in search time to the window indicates that it did not fit this schema and further indicates a differential processing of target and background that has been suggested previously (e.g., Neisser, 1967). Possibly, then, in odd-item search, the target is matched with a remembered representation, whereas the background form, though crucially affecting search speed, need not be identified in memory.

Dependence of Similarity Data on Method

Finally, the use of visual search to measure similarity deserves comment. The search task contrasts with the discrimination tasks, with which similarity is typically measured in animal subjects. In search, forms to be compared are simultaneously presented, and confusions might be attributed to perceptual processes. In the typical discrimination task, forms are learned one by one and confusions might arise in the comparison between immediate input and remembered representations. Comparable results from search and learning tasks would suggest that stimulus similarity is determined by processes common to both. One might then pursue with confidence a general psychophysics of form. If the methods gave distinctly different results, one might attempt to disentangle the relative contribution of perceptual and memory processes.

Some experiments with humans suggest that similarity among forms is strongly dependent on the method used to make the measurement. For example, Podgorny and Garner (1979) had subjects rate alphabetic letters for similarity and also discriminate the letters under speeded conditions. Both measures are related to similarity, yet their correlation was only .59. Results from other experiments suggest that rapid visual search may be dominated by relatively local analysis of simple features, and thus it also may yield rather special results (e.g., Julesz, 1981; Triesman & Gelade, 1980).

John Franklin and I (Blough & Franklin, 1985) compared two estimates of letter similarity in pigeons. One measure came from my earlier studies, in which birds learned individual target letters (D. Blough, 1982, 1985). The other measure came from a version of odd-item search, in which a small group of identical target letters was found on a large array of another letter. Rather to our surprise, the results from these quite different methods produced very similar results. The mean data matrices correlated .89, which was very high, especially in view of some high interbird differences that we observed. The results of cluster analysis and of multidimensional scaling for the two data sets, though differing in some details, provided no suggestion of systematic differences. These results suggest that tasks stressing memory for form may not introduce unique influences on similarity beyond those implicated in the odd-item search task.

OVERVIEW

This chapter has discussed how search speeds reflect the pairwise similarities of forms, how these similarities may define significant aspects of form, and how these aspects may jointly contribute to the similarity and categorization of forms. Let us briefly review these matters.

As to the search task, appropriately transformed search speed seems an effective psychophysical measure, but we have seen only hints about search itself. Like humans, pigeons seem more easily to find objects marked by the presence of a feature (e.g., a tail) than by the absence of a feature, and like humans, pigeons appear to bring memory more strongly to bear on a target than on a background. Still obscure are the computations through which the "differentness" of an odd item is extracted from the "sameness" of a background.

As to elements or features, when these are clearly dictated in the composition of stimuli, they reappear in the similarity structure of the data (experiments 1 and 2). But similarity structure can suggest unexpected features (experiment 3). Perhaps the use of pseudonatural stimuli may guide us toward critical aspects of real-world objects.

As to the processes of dimensional or featural combination, it is doubtful that anything as simple as an additive combination rule can explain the discrimination of form. Combination rules are relatively clear and explicit, however; by pushing them as far as possible and observing their points of failure, we may gradually approach more correct and complex solutions. For example, systematic data showing additivity (experiment 1) challenge more sophisticated models like those suggested in other chapters of this book.

This work was done with pigeons. The contributions of comparative work need not again be listed, but certain of them are conspicuous here. Pigeons and humans seem to face many similar adaptive problems in the use of visual information; by comparing their individual solutions, we may detect important generalities or see certain aspects of visual processing with particular clarity. Pigeons and humans need not see alphabetic letters in a similar way, but evidence suggests that they do (D. Blough, 1985). Does this mean that advanced nervous systems apply a common solution to the recognition of such forms? In contrast, pigeons and humans may differ in the high-speed processing of textures; evidence suggests that processing across a wide expanse of the visual field is carried out more completely in pigeons than in people (Blough & Franklin, 1985). Contrasting data on this matter could help to isolate and define a presumed early stage of processing in humans. Such hints as these, and systematic data, are more than enough to keep our pigeons employed for such of the future as is rashly called *foreseeable*.

ACKNOWLEDGMENT

This research was supported in part by NSF grant BNS-8025515.

REFERENCES

Beals, R., Krantz, D. H., & Tversky, A. (1968). Foundations of multidimensional scaling. *Psychological Review, 75,* 127–142.

Blough, D. S. (1972). Recognition by the pigeon of stimuli varying in two dimensions. *Journal of the Experimental Analysis of Behavior, 18,* 345–367.

Blough, D. S. (1977). Visual search in the pigeon: Hunt and peck method. *Science, 196,* 1013–1014.

Blough, D. S. (1979). Effects of the number and form of stimuli on visual search in the pigeon. *Journal of Experimental Psychology: Animal Behavior Processes, 5,* 211–223.

Blough, D. S. (1982). Pigeon recognition of letters of the alphabet. *Science, 218,* 397–398.

Blough, D. S. (1985). Discrimination of letters and random dot patterns by pigeons and humans. *Journal of Experimental Psychology: Animal Behavior Processes, 11,* 261–280.

Blough, D. S. (1986). Odd-item search by pigeons: Method, instrumentation, and uses. *Behavior Research Methods, Instruments, & Computers, 18,* 413–419.

Blough, D. S. (1988). Quantitative relations between visual search speed and target-distractor similarity. *Perception & Psychophysics, 43,* 57–71.

Blough, D. S., & Franklin, J. J. (1985). Pigeon discrimination of letters and other forms in texture displays. *Perception & Psychophysics, 38,* 523–532.

Blough, P. M. (1984). Visual search in pigeons: Effects of memory set size and display variables. *Perception & Psychophysics, 35,* 344–352.

Clauson, H. D., Izatt, E. J., & Shimp, C. P. (1985). An infrared system for the detection of a pigeon's pecks at alphanumeric characters on a TV screen: The dependency of letter detection on the predictability of one letter by another. *Journal of the Experimental Analysis of Behavior, 43,* 257–264.

D'Amato, M. R., Salmon, D. P., & Columbo, M. (1985). Extent and limits of the matching concept in monkeys (Cebus apella). *Journal of Experimental Psychology: Animal Behavior Processes, 11,* 35–41.

Garner, W. R. (1978). Aspects of a stimulus: Features, dimensions, and configurations. In E. Rosch & B. B. Lloyd (Eds.), *Cognition and categorization* (pp. 99–133). Hillsdale, NJ: Lawrence Erlbaum Associates.

Garner, W. R., & Felfoldy, G. L. (1970). Integrality of stimulus dimensions in various types of information processing. *Cognitive Psychology, 1,* 224–241.

Hearst, E., & Jenkins, H. M. (1974). *Sign-tracking: The stimulus-reinforcer relation and directed action.* Austin, TX: Psychonomic Society.

Julesz, B. (1981). Textons, the elements of texture perception, and their interactions. *Nature, 290,* 91–97.

Julesz, B., & Schumer, R. A. (1981). Early visual perception. *Annual Review of Psychology, 32,* 575–627.

Kubovy, M., & Pomerantz, J. R. (Eds.). (1981). *Perceptual organization.* Hillsdale, NJ: Lawrence Erlbaum Associates.

Neisser, U. (1967). *Cognitive psychology.* New York: Appleton-Century-Crofts.

Podgorny, P., & Garner, W. R. (1979). Reaction time as a measure of inter and intraobject visual similarity: Letters of the alphabet. *Perception & Psychophysics, 26,* 37–52.

Real, P. G., Iannazzi, R., Kamil, A. C., & Heinrich, B. (1984). Discrimination and generalization of leaf damage by blue jays (Cyanocitta cristata). *Animal Learning & Behavior, 12,* 202–208.

Sands, S. F., Lincoln, C. E., & Wright, A. A. (1982). Pictorial similarity judgments and the organization of visual memory in the rhesus monkey. *Journal of Experimental Psychology: General, 111,* 369–389.

SAS Institute (1982). *SAS System* (computer program). Cary, NC: Author.

Shepard, R. N. (1980). Multidimensional scaling, tree-fitting, and clustering. *Science, 210,* 390–398.

Treisman, A. M., & Gelade, G. (1980). A feature-integration theory of attention. *Cognition Psychology, 12,* 97–136.

Treisman, A. M., & Souther, J. (1985). Search asymmetry: A diagnostic for preattentive processing of separable features. *Journal of Experimental Psychology: General, 114,* 285–310.

Tversky, A. (1977). Features of similarity. *Psychological Review, 84,* 327–352.

Wiener-Ehrlich, W. K. (1978). Dimensional and metric structures in multidimensional stimuli. *Perception & Psychophysics, 24,* 399–414.

8 Shape Constancy in the Pigeon: The Perspective Transformations Decomposed

John Cerella
Veterans Administration Outpatient Clinic
Boston, Massachusetts

Previous Work

The transformations of perspective can be regarded as an unwanted source of target variability from the point of view of an image-processing device required to detect a particular object from a succession of objects. Such a device might well recognize the group structure of the set of all possible projections of the target and rely on an inverse, normalizing transform to identify any given projection. This would allow the valid line projections of a target to be distinguished from distorted line projections possessing the same set of local features.

In 1977, I published three experiments proporting to demonstrate that pigeons were incapable of this distinction. Given the inaccessibility of the original publication, I review that work here. In the first experiment a pair of white Carneaux pigeons was first trained to discriminate a square from random quadrilaterals. Figures were generated by an on-line, time-shared computer and presented one at a time for a brief period on a small TV screen within a conventional operant chamber. A burst of pecks to the screen while a square was showing was reinforced by a 2-second presentation of grain. Responding to a nonsquare served only to prolong the image on the screen. Over the course of several months, the discrimination ratio (that is, the probability of a response burst to a square divided by the total probability of response bursts) rose from 0.50 to over 0.75 per session for both birds.

Random polar projections of a cube at a fixed distance from the viewer (as in Fig. 8.1a) then replaced squares as positive patterns. Negative patterns were derived from positive projections by displacing each vertex of a projection a small random amount to break up the mathematics and the ''Gestalt'' of the

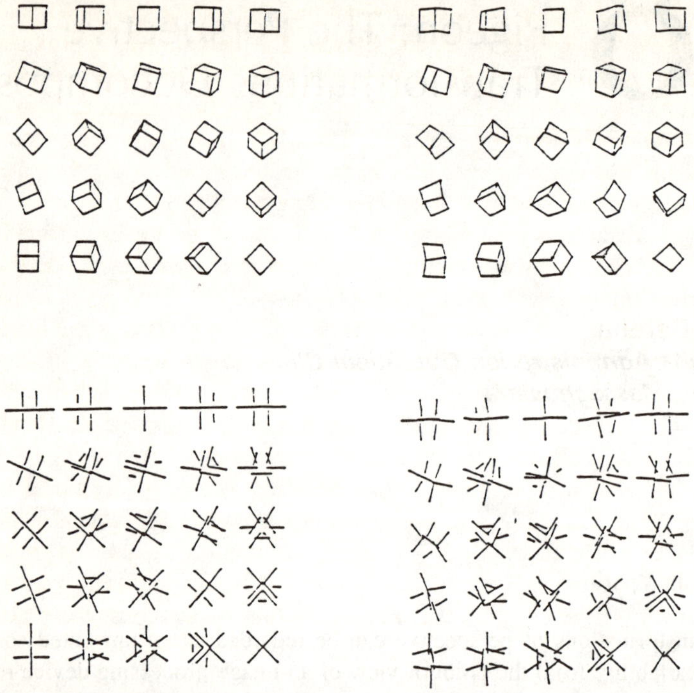

FIG. 8.1. Pigeons were unable to discriminate cubes (A: upper left) from distorted cubes (B: upper right). Twenty-five of 9,261 possible projections are illustrated. Subsets of nearly congruent views could be learned, but this was equally true of the abstract patterns in the lower half of the figure. Patterns in the lower left (C) "correspond" to cubes but are not three dimensional. Patterns in the lower right (D) correspond to distorted cubes.

cube. In this way an unlimited population of "flat," "two-dimensional" patterns was produced having the same distribution of graphic elements (number of lines, vertices, and regions) as the population of positives (Fig. 8.1b).

Discrimination ratios on this problem dropped to 0.55 and 0.60, respectively, and remained at that level for over 150 sessions. These values did not differ from chance for any single session (Chi-square test). Finally, a return to quadrilaterals restored the discrimination ratio to about 0.80 for both birds. The pigeons therefore were able to discriminate a single projection of the cube (a square) but were unable to distinguish the class of all such projections despite extensive training.

Given that a single projection could be learned, I conceived the idea of introducing additional projections incrementally by progressively extending the angle through which the cube was allowed to rotate. In a second experiment the positive class was developed gradually in this manner, in an attempt to facilitate learning.

Due to symmetry, all possible views of the cube can be generated by passing it through 90° around the X, Y, and Z axes. This domain was approximated by the 21 discrete values 0°, ±10°, ±14°, ±18°, ±22°, ±26°, ±30°, ±34°, ±38°, ±42°, ±46°. Completely random projections result from drawing X, Y, and Z rotations randomly from these values. There are 21 × 21 × 21 = 9261 such views—this was the set of projections the pigeons failed to discriminate in the first study. In the second experiment, the initial stimulus set consisted of a single projection, given by a rotation of +46°, +46°, +46° on the X, Y, and Z axes, respectively. The cube was thereby oriented with a corner toward the viewer. The second stimulus set consisted of the 2 × 2 × 2 = 8 views given by rotations of either +46° or −46° on each of the axes. The third set comprised rotations of the ±46° or ±42° (4 × 4 × 4 = 64 views). The fourth set, of ±46°, ±42°, or ±38° (6 × 6 × 6 = 216 views), the fifth set of ±46°, ±42°, ±38°, or ±34° (8 × 8 × 8 = 512 views), and so on to the twelfth set that allowed all 21 rotations. Throughout the sequence negative patterns were derived from the complete set of positive projections—the population of negatives was never restricted.

Four White Carneaux pigeons were introduced to this sequence. A given stimulus set was enforced until a discrimination ration equal to or greater than 0.70 was achieved over 3 consecutive sessions. The pigeon was then advanced to the next set. If the criterion was not met within 30 sessions, the pigeon was discontinued.

All four pigeons learned the first 7 steps corresponding to a cube rotated ±24° or less around its starting position. None of the pigeons, however, learned the complete set of rotations (that is ±46° or less around the starting position). As the cube assumed more divergent orientations, the problem grew increasingly difficult; somewhere between ±24° and ±46° the performance of each bird fell to and remained at the level of chance. So, even when insinuated gradually, the complete set of projections of a cube remained beyond the discriminative capacity of the pigeon.

One might conjecture that the pigeons' progress through this graded problem sequence was nonetheless more rapid and more extended than it might have been given a sequence of purely abstract patterns of comparable complexity. In a third experiment, I constructed such a sequence. Abstract patterns were derived from the projections of a cube by rotating each of the line segments of a projection 90° around its midpoint. Because this transformation is performed on the two-dimensional projection, and not the three-dimensional cube, the resulting line patterns no longer represent any real object; collectively, however, they convey the same amount of information as the original projections and so are mathematically of an equivalent complexity. Negative patterns were derived from distorted cube projections by the same line rotation (Fig. 8.1c, d).

A different group of four White Carneaux pigeons was exposed to a sequence of abstract stimuli sets in exact parallel to the sequence of projections of experiment 2. Once again, all four pigeons learned the early stimulus sets, whereas fewer and fewer learned the later sets. The fall-off in accuracy and acquisition time as the stimulus sets advanced was the same in experiment 3 as in experiment 2. Thus, the abstract and the representational patterns did not differ in their psychological complexity for the pigeon.

Experiment 1 showed that pigeons were insensitive to the group structure of the set of projections of a three-dimensional object being unable to "invert" or identify arbitrary projections of that object. Experiments 2 and 3 allowed that conclusion to be strengthened. Granted that pigeons could not realize the complete projection group of a simple object, they could, nonetheless, learn a small subset of that group. What has been shown is that even this capacity was based on generalized "resemblance detection" rather than specific perspective processing, in that a small subset of an abstract pattern set could be learned as readily. I was unable to detect any sensitivity whatsoever to the perspective transformation as such.

Gross Transformations

Stimulated in part by Hollard and Delius' (1982) remarkable findings to the contrary, I have returned to the question of perspective invariance in the pigeon and report my more recent results. The 1977 study confounded X-, Y-, and Z-rotations and did not treat dilations or shifts. In the following experiments a prototype was established, and each of these transformations was applied separately.

An alphabet was devised on a 16-segment skeleton, with the letters F, G, J, R reserved as targets, and the remaining 22 letters serving as negatives. The letters were unusual in having a 1:1 aspect ratio, i.e., a square rather than a rectangular format, so that rotations would not be cued by a shift in the height–width ratio. Ten transformations were defined (Fig. 8.2), six rotations around various axes,

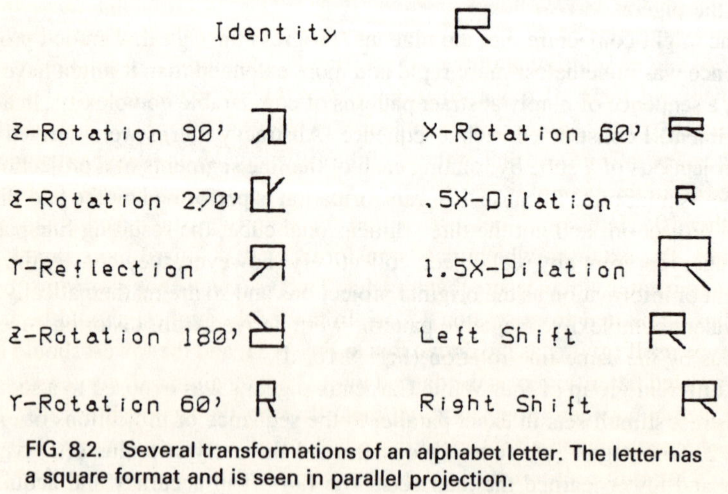

FIG. 8.2. Several transformations of an alphabet letter. The letter has a square format and is seen in parallel projection.

two dilations, and two translations. The transformations were crudely scaled so that the product would be more or less "half on and half off" the prototype when superimposed.

As in the 1977 study, computer-generated images were delivered to a small TV mounted at eye level in the wall of an operant chamber; responses were collected directly from the screen that was floated on a microswitch. Unlike the 1977 study, the new experiments used the extended trial, multiple variable interval/extinction schedule more typical of concept-formation studies, as refined by Vaughan and Greene (1984). The procedure encouraged a high rate of pecking to positives, and a low rate of pecking to negatives; data were recorded over the first 10 seconds of a trial, before schedule contingencies were manifest.

The first experiment followed the logic of the earlier study. A prototype was established and then transforms of the prototype were introduced as an extension of the positive class. The response to each transform was tracked separately, in expectation that the rate of acquisition would index the degree of invariance in each case. For example, transfer to the shifted prototype might be immediate, whereas transfer to the upside down prototype might take several training sessions to establish. Although attractive in theory, this design failed in practice, as we see next.

Four homing pigeons were employed, each assigned one of F, G, J, or R as a target. Two birds saw large letters ($\frac{1}{2} \times \frac{1}{2}$ inch, 12.7mm × 12.7mm), and two birds saw small letters ($\frac{1}{8} \times \frac{1}{8}$ inch, 3.18mm × 3.18mm). The target was first contrasted with the 22 negative letters. The discrimination was acquired easily by all birds over 14 training sessions, as shown by the dotted line in the acquisition curves (Fig. 8.3).

On the 15th session, 10 transformations of the target were introduced as additional positives. Negative stimuli were multiplied at the same time, being subject to the same transformations. A 12th positive was also introduced, one of the four possible targets not seen by the pigeon. The full stimulus ensemble for one pigeon consisted of the original target F, 10 transformations of F, also positive, and an additional unrelated positive R, plus the 22 nontarget letters as negatives and their 10 transformations (22 × 11 = 242 total), also negative.

Starting on the 15th session the pigeon was required to learn 11 new positives. Let us accept the premise that of these 11 the only way the unrelated letter could be learned (R in the example) was disjunctively, i.e., as a separate case. Not so with the other 10. Here some measure of transfer from the established positive was possible. The extent to which these transforms were acquired more rapidly than the random letter will measure that transfer. Twelve concurrent acquisition curves can thus be plotted from session 15 on. In theory the highest should be for the prototype itself (giving as the dotted line in Fig. 8.3), and the lowest should be for the unrelated letter (given as the dashed line). The remaining 10 curves should be positioned in between (given as solid lines). In fact, the figure shows that the 12 curves for each pigeon were superimposed. Evidently, the sudden introduction of 253 novel stimuli confused the birds to the extent that even the original discrimina-

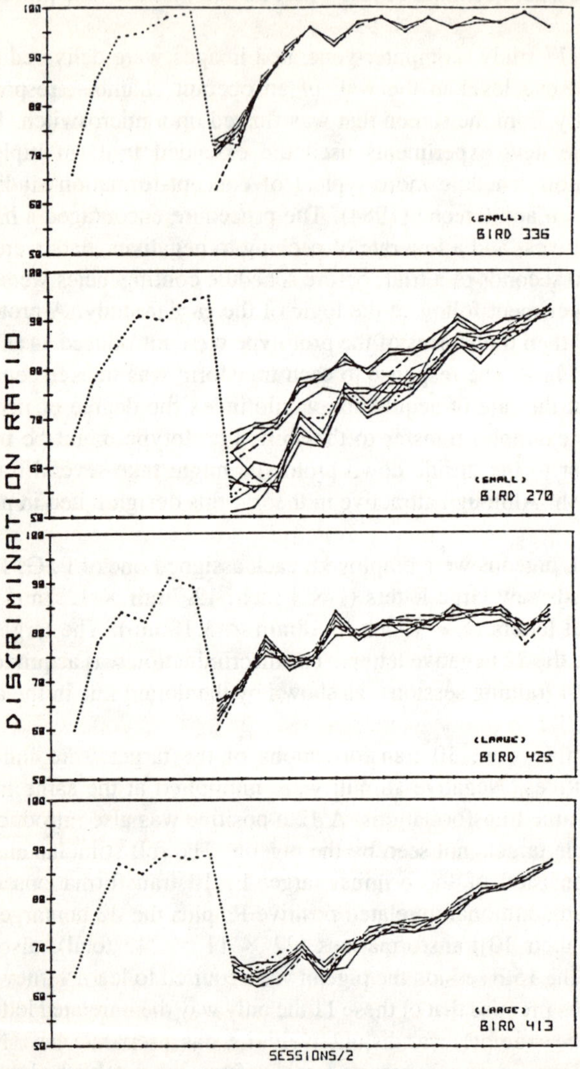

FIG. 8.3. Sessions 1–14: Birds readily learned to distinguish 1 positive, an alphabet letter, from 22 other letters (dotted line). Sessions 15–45: The 10 transforms of Fig. 8.2 were added to the positive set (solid lines), together with a random letter (dashed line). All 12 patterns were learned from scratch, with no transfer from prototype to probes.

tion was lost. All 12 positives were learned anew, and the coincidence of the curves suggests they were learned as 12 discrete cases. The experiment failed to yield any convincing information about perceptual invariances.

A second experiment was undertaken, reversing the logic of the first. In the second experiment responding was initially established to the prototype alone. Then the 242 negatives were introduced, as well as the 10 prototype transforms. Rather than being designated positive, the latter were inserted in infrequent, brief, unreinforced, "probe" trials. The effect of invariance in this experiment would be the more gradual extinction of a probe in comparison to the explicit negatives, rather than the more rapid acquisition expected in the first experiment.

The procedure was applied to eight homing pigeons, two on each target. Four of the pigeons then repeated the procedure on a different target, permitting differences between targets and differences between observers to be separated. Only small letters ($\frac{1}{8} \times \frac{1}{8}$ inch, 3.18mm $\times$ 3.18mm) were used.

Figure 8.4a presents the set of extinction gradients collected from 1 bird viewing 1 target over the 18 training sessions that followed the introduction of the test stimuli (responding to the prototype was pre-established). Plotted is the "confusion ratio" given by the response to a test stimulus divided by the response to the prototype. On this measure, response to the prototype itself would show as a horizontal line at 1.00; this line is omitted. Eleven curves are shown, one for each of the 10 transforms (labeled at both ends) and one for the 242 negatives combined (labeled X).

The curves are tangled; worse, there are 11 other sets, also tangled. Sifting the sense from the tangles required considerable abstraction; be warned that other analyses may lead to other conclusions than the ones to follow. To me, at any rate, it seemed that two aspects of a bird's response to a probe must be noted, the initial level of response, and the fall-off in response as training progressed. These tendencies were more easily discerned when the extinction gradients were smoothed. The curves of Fig. 8.4a have been fitted by power functions in Fig. 8.4b. The average goodness of fit over the 12 replications $\times$ 11 probes = 132 gradients was $r = .67$. The fitted functions followed the trends of the originals fairly well (the residuals were random).

The gradients can now be assigned to "transfer classes" by a cluster analysis. A power function is linear in log–log coordinates, and the two parameters of the log–linear function perfectly express the crucial aspects of the data, the Y-intercept giving the initial level and the slope giving the rate of fall-off. Figure 8.4c gives the parameter space of the functions of Fig. 8.4b. Curves in Fig. 8.4b map onto points in Fig. 8.4c, with the ordinate fixed by the intercept value and the abscissa by the slope value.

At this stage the "K-means" clustering algorithm was applied, with two clusters predefined, one by the point representing positives and a second by the point representing negatives. The action of the algorithm was to assign the remaining points to the existing clusters, or to a newly created third or fourth

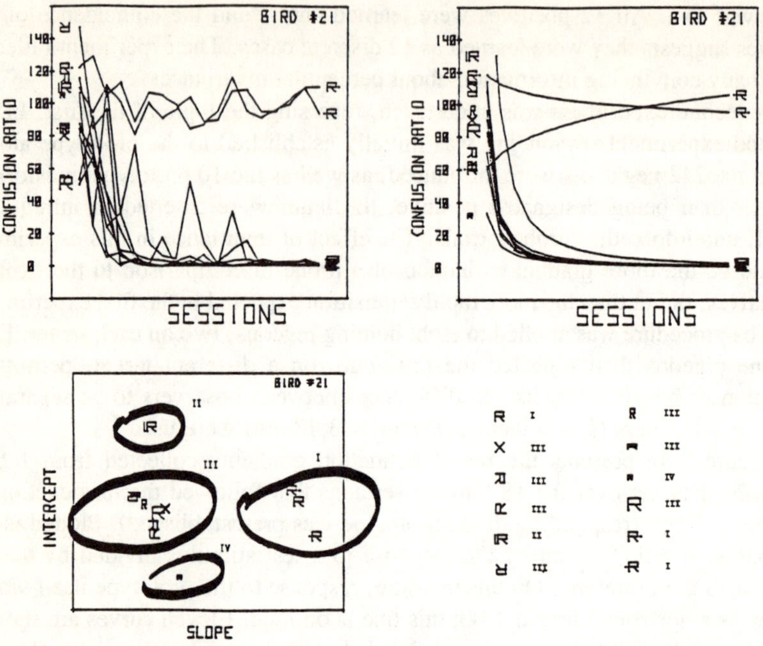

FIG. 8.4. A (upper left): Response levels to the 10 probes of Fig. 8.2 over the course of acquisition of a discrimination between 1 positive letter and 22 negative letters. Response to the positive would plot as a horizontal line at 100 (not shown). Average response to negatives is plotted by the curve marked X at each end. B (upper right): Power functions have been fitted to the 11 curves of A. C (lower left): The parameter space of B (functions in B plot as points in C). A clustering algorithm has been applied, that has returned four clusters. D (lower right): Cluster membership as given by C. Cluster I was defined as including the positive; Cluster III was defined as including the negatives. With respect to response levels, Cluster IV was "supernegative" and Cluster II was "semipositive."

cluster, depending on the goodness of fit. Finally, the resultant clusters were identified as class I, II, III, or IV. Class I points included the positive by definition; the points all reflected high initial values and sustained responding over subsequent sessions. Class II points showed initial values comparable to the positive, but subsequent fall-off to final values comparable to the negatives. Class III points included the negatives by definition; the points showed moderate initial values and subsequent fall-off. Class IV points showed low initial values, and because of the cellar effect, little subsequent decay. These assignments are listed in tabular form in Fig. 8.4d. There are 12 entries, the first 2 (positives and negatives) being fixed by the analysis. The remaining 10 give the transfer class assigned to each probe and constitute the "invariance profile" for a particular pigeon and target.

We can at last pose substantive questions. First, target differences and observer differences can be compared. In four cases two naive birds were trained on the same target. These four pairs of invariance profiles were on the average less than a half transfer class apart, indicating a high level of agreement between birds. In four cases the same bird was trained on two targets. These four pairs of profiles were on the average nearly one transfer class apart. Thus, target differences were responsible for two times as much variation as observer differences.

Observer differences were averaged out. Each target was viewed by three pigeons, and Fig. 8.5 gives the transfer class assigned to each probe by the

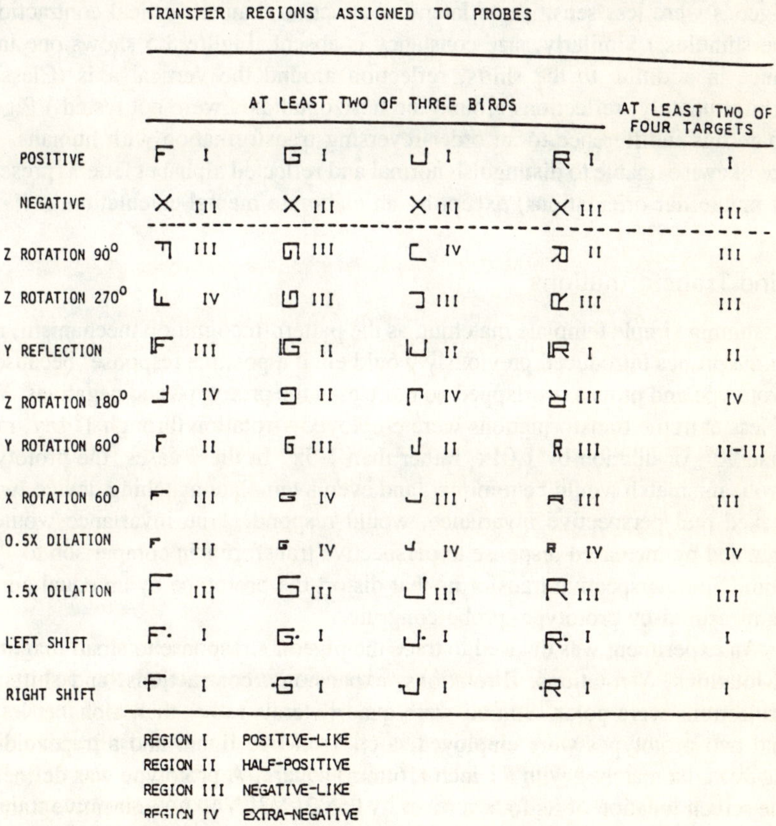

FIG. 8.5. Results of the curve-fitting and -clustering analysis illustrated in Fig. 8.4 for one bird, applied to all 12 bird-target combinations. First four columns give cluster membership by target; last column sums over all targets.

majority of the three. The resulting four profiles show a number of differences, due to the identity of the target; they are not considered here. Instead, the common tendencies have been summarized in the right-most column of the table. These entries are of considerable interest. The most striking outcomes are the last two, left shift and right shift; these data indicate complete shift invariance (Class I). At the opposite extreme, the 0.5x dilation and Z180 rotation show negative transfer (Class IV). These probes were more readily discriminated from the target than were the negatives. The case of the 0.5x dilation may be an artifact of the limited screen resolution. The segments of these minute stimuli tended to coalesce and were perhaps coded as "blobs" rather than "patterns." The pigeons rapidly learned that blobs were not reinforced.

No orientation constancy is apparent in Fig. 8.5, with the exception of Y60, which shows as Class II for two targets. (X60 forms an interesting contrast: The pigeons were less sensitive to lateral contraction than to vertical contraction of the stimulus.) Similarly, size constancy is absent. Figure 8.5 shows one invariance in addition to the shifts, reflection around the vertical axis (Class II). (Unfortunately, reflections around the horizontal axis were not tested.) Pigeons share this indifference to an order-reversing transformation with humans, who are likewise unable to distinguish normal and reflected alphabet letters (presented in unfamiliar orientations) except by an elaborate mental calculation.

Fine Transformations

Assuming simple template matching as the pattern-recognition mechanism, none of the probes introduced previously would elicit a positive response, because the prototype and probes overlapped no more than the prototype and negatives. What if less extreme transformations were employed—rotation through 1° say, rather than 90°, or dilation by 1.01x, rather than 1.5x? In these cases, the prototype–probe mismatch would be minimal, and even a template matching device, which lacked real perspective invariance, would respond. True invariance would be signaled by increased response to perspective transforms in comparison to "random," nonperspective transforms that distort the prototype by an equal amount as measured by prototype–probe congruity.

An experiment was devised to trace the pigeon's response to small magnitude X-rotations, Y-rotations, Z-rotations, expansions, contractions, and shifts; the projections were polar. Stimuli were quadrilaterals rather than alphabet letters, and two prototypes were employed, a chevron-like figure and a trapezoid-like figure, area matched with a $\frac{5}{8}$ inch (16 mm) square. A prototype was defined by the screen location of its four vertices, V1, V2, V3, V4. A perspective transformation has the effect of displacing each vertex a small distance, d1, d2, d3, d4, from its home position. The sum of the four displacements may be adopted as a measure of the "Euclidean" distortion induced by the transformation.

Five levels of each of the perspective transformations were selected—five X-rotations, five dilations, etc—to induce increasing amounts of prototype distortion in a graded sequence. In addition to the six perspective transformations, a random transformation was constructed that deformed the prototype in a nonsystematic way. The five levels of all seven transformations were equated, so that the magnitude of the pigeon's response to each transformation type could be compared on a common distortion axis.

The procedure followed the preceding experiment, in that a positive–negative discrimination was established, and response to occasional unreinforced probes was measured. Four birds were used, two trained on the chevron and two on the trapezoid. Negatives were irregular, randomly generated quadrilaterals. Data were collected in a steady state, rather than during acquisition or extinction.

Data, accumulated over 28 sessions after a steady state had been achieved, are shown in Fig. 8.6. There are 7 panels, 1 for each transformation type. Each panel shows the response levels of the 4 birds across the 5 levels of the transformation, as a percentage of the response to the untransformed prototype that marks the left end of the distortion axis (and is therefore fixed at 100%). Inserted in each panel is also the complete set of probes, superimposed on the prototype. Of course, these figures were seen individually, not superimposed, and ought by right to be positioned along the distortion axis. Further, only the chevron-family is given, the trapezoid-family is not illustrated.

In most cases (4 birds × 7 transform types = 28 curves), responding decreased as distortion increased. The critical question is how rapid was the fall-off? The comparison must be with the bottom panel, the "random" transformation. These curves reflect the resolution of the pattern-matching operation. If responding in any of the upper six panels was elevated, some mechanism beyond simple pattern matching must have intervened to "cancel" what the bottom panel demonstrates is a detectable prototype–probe discrepancy.

It appears that the six upper panels divide clearly into two groups. The three on the left show response fall-off at least equal to the bottom panel. Thus, for X-rotations, Y-rotations, and Z-rotations, no variance is manifest. The three panels on the right show sustained responding in the face of extreme geometric disparities. Thus, invariance is demonstrated for expansions, contractions, and shifts.

How do these results on fine transformations mesh with those on gross transformations? Both studies show position constancy, and neither show orientation constancy. The discrepancy in size constancy may be only at first sight. The gradients in Fig. 8.6 fall off toward the extreme levels of expansion (x1.4) and contraction (x0.6). The levels of the previous experiment were yet more extreme, x1.5 and x0.5. Extrapolating Fig. 8.6 suggests that at those values responding would more or less drop to zero. The results may therefore be complimentary, constancy is seen over small size changes but not over large.

The two studies, then, are in fair agreement. One tendency that shows in the first and not in the second is the elevated response to Y-rotations. This may be

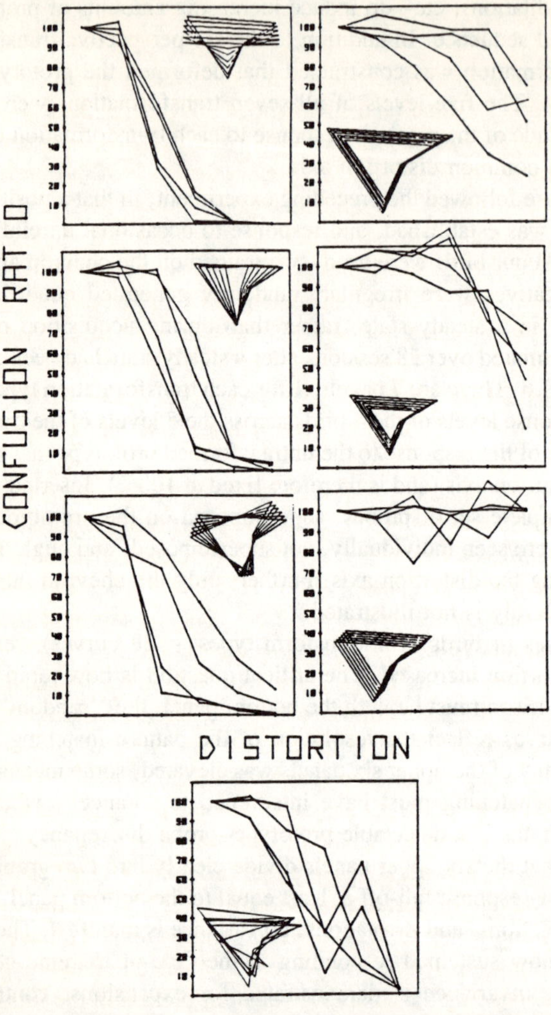

FIG. 8.6. Steady-state generalization gradients to 7 series of 5 calibrated probes. Probes are shown superimposed in the inserts but were viewed individually. The bottom panel shows random deformations. The perspective deformations in the three left-hand panels elicited about the same level of response as the random deformations. The perspective deformations in the three right-hand panels elicited more response.

due to the nature of the projection, parallel in the first case and polar in the second. It is on this basis that we have associated the effect with "lateral contraction" rather than "rotation," but it is also possible that the targets rather than the projection makes the difference.

Three-Dimensional Transformations

Hidden features of a solid object (e.g., the eyes, ears, and hair of a head) are exposed as it turns. "Rotational invariance" may here name a disposition to associate such orderly feature sequences (hair/ear/eye/nose/eye/ear) more readily than haphazard sequences (eye/hair/ear/nose/eye) of the same features. In this experiment two pattern recognition problems were devised to test this disposition in the pigeon. In one problem, the target was a hexagonal prism, with the letters A, B, C, D, E, and G inscribed on its six faces. The prism was seen from the side and stepped in 60° increments, so that exactly three faces were visible at a time. Consequently, there were six, and only six, projections ABC, BCD, CDE, DEG, EGA, GAB. In the other problem, six random permutations of the same letter set constituted the positives, ADE, EDG, GBD, CEA, BAC, and GBC. Collectively, the random permutations describe no three-dimensional structure.

One group of four pigeons was required to learn the set of cyclic permutations, another group was required to learn the set of random permutations. In the first phase of training, the negatives for both problems consisted of letter triplets drawn randomly from the alphabet letters H–Z ($19 * 18 * 17 = 5814$ possible stimuli). In both cases positives and negatives could be distinguished on the basis of local features, and neither group of pigeons should have difficulty.

In the second phase of training, a new set of negatives replaced the old. The new negatives were the 108 permutations of the letters A, B, C, D, E, G, taken three at a time, which remain after the 12 permutations already serving as positives (six in each problem) are removed.

In the second phase, positives and negatives can no longer be separated at the local feature level. A device capable of detecting only local features could not separate the positives and negatives of either problem. The six random permutations could, however, be learned individually, via six "global" templates, a solution equally applicable to the six cyclic permutations. To the extent that the cyclic permutations are learned more readily than the random permutations, pigeons may be encoding the adjacency relations present in the former and not in the latter. Thus, a divergence in the acquisition curves for the two problems would constitute evidence for rotational invariance as defined earlier.

The average acquisition curves for the 2 groups of 4 pigeons are given in Fig. 8.7. The first segments of the curves depict phase 1; as expected, error scores for both groups dropped rapidly over 8 sessions. The 36 sessions of phase 2 are more interesting. Both groups were forced to chance (about 20 errors per session) at the outset. Thus, adjacency relations had not been encoded during phase 1. Further, error rates for both groups declined far more gradually—clearly the phase-2 discrimination was more difficult than the phase-1. Nevertheless, the discrimination was acquired more rapidly by the cyclic group. The internal structure of the stimuli was evidently perceived and used to reject specious "instances," tantamount to rotational invariance.

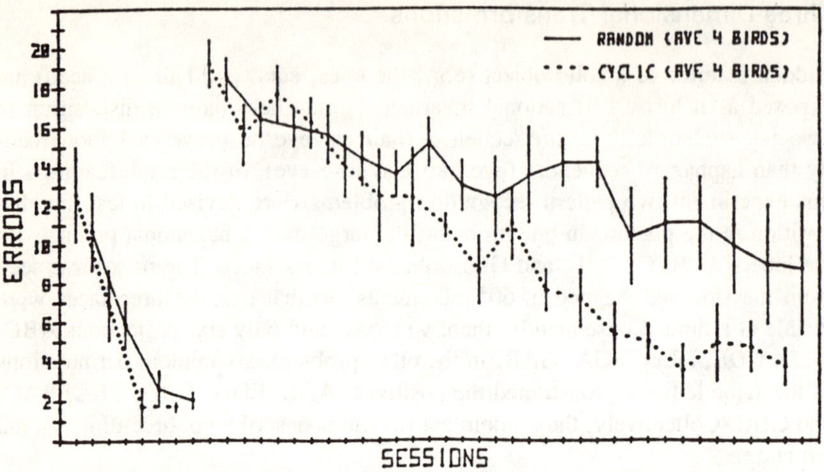

FIG. 8.7. Pigeons were required to learn a set of six-letter triplets. Positive triplets were readily separated from negative triplets composed of different alphabet letters (first eight sessions). Separating the positive triplets from negative triplets composed from the same letters was more difficult (sessions 9–45). Cyclic permutations were easier to learn than random permutations in the latter condition. Pigeons may have cued on the repeated duples present in the cyclic permutations.

Unfortunately, the stimuli also allow a solution at the level of duples. We are assured that in phase one single letters were encoded because the discrimination collapsed at the start of phase two. Cyclic birds may then have noticed that certain letter pairs were more likely to occur in positives, e.g., the pair AB occurred in two and so constructed a template of this duple. The positive ABC may thus have been encoded as one duple, AB, one singlet, C, and one adjacency relation AB + C. There was no repeating duples among the noncyclic positives, and, thus, there would be no advantage in coding duples over three singlets and two adjacency relations, A + B + C. It is possible, therefore, that cyclic birds exploited only the 2D and not the 3D regularity of the letter strings.

Translations

Surely the pigeon can recognize an X anywhere on the screen, perhaps by a head movement that recenters the target. True shift invariance implies target identification at eccentric locations without a redirection of gaze. In a final experiment, identification latencies were traced as a function of target eccentricity in an effort to separate these alternatives.

The stimulus for a trial was a single $\frac{1}{16}$ inch (1.59mm) letter, either X or 0. A single peck or 3 seconds without a peck terminated the trial. If X, a peck tripped

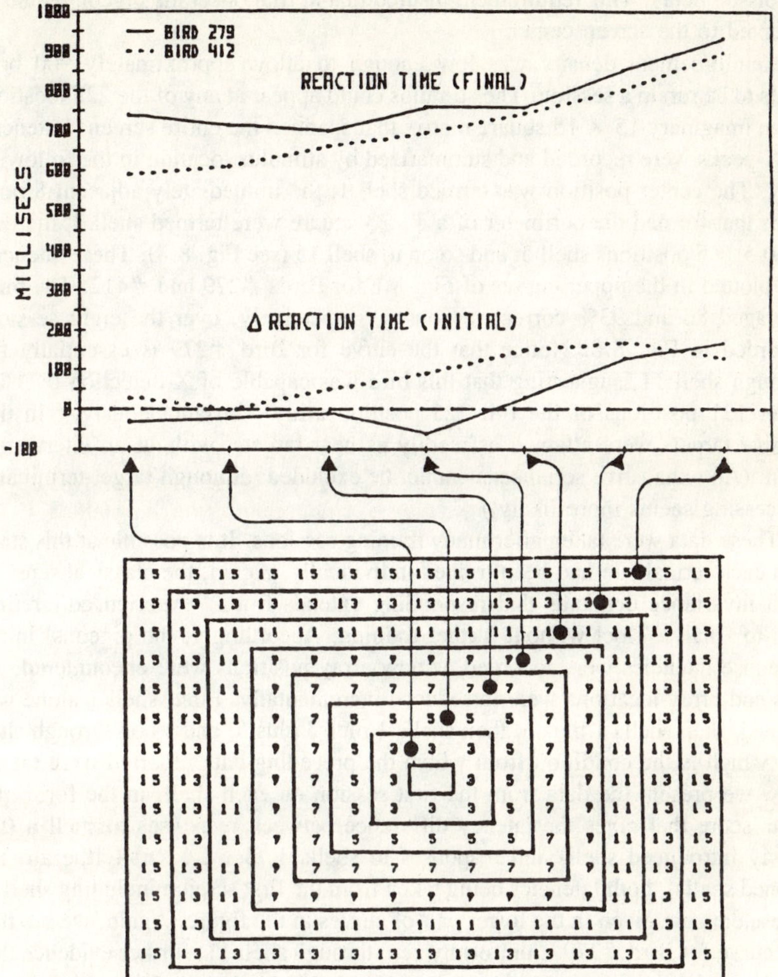

FIG. 8.8. Upper panel: time taken to respond to an X in any one of the seven shells diagrammed in the lower panel. Trials were tripped by striking a * occurring in the centermost position. Pigeons rejected an 0 with about 90% accuracy throughout the procedure. The lower pair of curves plot the difference in latency to a target appearing in shell n and targets appearing in shells 1 through n−1. Lower curves are taken from the pigeons' first exposure to shell n.

the hopper with probability .33; if 0, a peck produced a time-out. Trials began with an * at center screen. A peck to the * triggered the stimulus letter following a Poisson delay. This requirement insured that at trial onset the pigeon would be oriented to the screen center.

Reinforcement density was low enough to allow approximately 900 brief trials to be run in a session. The stimulus could appear at any of the 225 locations of an imaginary 15 × 15 square matrix that spanned the entire screen. Latencies of X-pecks were recorded and summarized by stimulus location in the following way. The center position was termed shell 1; the immediately adjacent 8 positions that formed the perimeter of a 3 × 3 square were termed shell 3; the next most 5 × 5 positions shell 5; and so on to shell 15 (see Fig. 8.8). These latencies are plotted in the upper curves of Fig. 8.8 for Birds #279 and #412. The birds averaged 86 and 93% correct responses, respectively, over the eight sessions recorded in Fig. 8.8. Notice that the curve for Bird #279 is essentially flat through shell 11, suggesting that this bird was capable of X-detection over the inner 121 positions of the full 225 position array "instantaneously," in that remote targets were attained as readily as near targets, without an intervening scan. (An exhaustive serial scan cannot be excluded, although target-terminated processing seems more likely.)

These data were taken after many training sessions. It is possible at this stage that each retinal site had been trained individually. Again, the classical sense of shift invariance demands that responding extends from "experienced" retinal loci to "naive" loci without further training. Accordingly, the pigeons' initial response latencies were recorded as new array positions were encountered. To this end array locations were introduced incrementally. First, shell 1 alone was trained, then shells 1 plus 3, then shells 1 plus 3 plus 5, and so on through shell 15, which is the condition from which the preceding data reported were taken. Now we present the data from the first session on each shell, in the form of a delta score that gives the latency difference between responses to shell n (the newly introduced shell) and responses to shells 1, 3, . . . , n–1 (the already trained shells), both latencies being taken from the first session including shell n. These data are given in the lower pair of curves in the figure. Again, we see that the curve for Bird #279 is uniformly zero through shell 11, further evidence that this animal enjoys true position constancy. At the same time Bird #412 found the translation problem manageable (with an accuracy of 93%), but at the cost of an additional processing time, time that was perhaps spent realigning the image.

Conclusions

The transformations of perspective are several, but they pose a common problem, critical patterns are deformed and to be recognized the deformation must be reversed by the observer. Our research has confronted the pigeon with each of the perspective transformations in turn and has recorded a varied pattern of success and failure. Success was seen with shifts, small dilations, mirror reversals, and lateral contractions. In each case there was at least an initial tendency to treat the deformed probe as equivalent to the prototype. Failure was seen with rotations around the X, Y, and Z axes. In these cases probes were treated as equivalent to negatives.

These findings are consistent with several others in the literature. The pigeons' failure to respond to letters rotated in the projection plane was found by a student of Herrnstein's (Cosmides, 1979) and corroborates my 1977 study on cubes. The pigeons' success with shifts was presaged by the Bloughs' work on letter detection. A student of Blough's has systematically documented the pigeons' "confusion" of reflected images (Todrin & Blough, 1983). Vaughan and Greene (1984) anticipated both the success with Y reflection and the failure with Z180 rotation. Lombardi and Delius (this volume) have demonstrated size constancy over a range (greater than x0.5 but less than x2.0) more or less coincident with mine (greater than x0.5 but less than x1.5).

In contrast to these findings is the remarkable report of Hollard and Delius (1982) of successful transfer from a prototype to a probe rotated in the projection plane. The negatives in their procedure were mirror reflections. As I understand it, this result is puzzling. On the face of it, two tendencies of the pigeon apparent in the present research are contradicted, the tendency to treat a mirror image as the same as the original, and the tendency to treat a rotated image as different from the original.

Perhaps, as Delius has suggested in his chapter, these tendencies can be countered with training. Indeed, my data (Fig. 8.4) and Todrin and Blough's show that the discrimination of mirror-image pairs is possible, but many trials were required per pair. In one attempt to counter the second tendency, to train the pigeon to treat as equivalent the rotations of a prototype, no number of trials availed (Cerella, 1977; it is unfortunate that a second attempt, depicted in Fig. 8.3, was a procedural failure). Yet in their final state, the pigeons of Hollard and Delius responded correctly to rotated mirror-image pairs on their first trial. Clearly, there is a pressing need to bridge the gap between fixed-target classification procedures and variable target matching-to-sample procedures. In the meantime, we must allow that the present studies may not have contacted the true competence of the pigeon.

Setting this unsettling reservation aside, we might consider in closing what mechanism is implicated by the present pattern of results. Current theorists attempt to derive the pattern invariances from the geometry of the retinal–cortical mapping. A variety of polar, logarithmic, and frequency transforms have been proposed, leading to an equal variety of predicted invariances (Cavanagh, 1978; Pollen & Ronner, 1982; Schwartz, 1980). One postulate has been common to all the models, a polar to rectangular conversion enacted in passing from the retinal to the cortical surface. Coupled with a secondary, shift-invariant transform, this is sufficient to invert rotations in the image plane. Signs of this inversion are totally lacking in the behavioral data presented here. If the data hold, a fundamental difference between the avian and mammalian visual systems is implicated. On the other hand, these theorists leave at our disposal neurally plausible implementations of logarithmic and frequency transforms sufficient to account for the size and position constancies (respectively) that are displayed by the pigeon.

In their chapter, Heinemann and Chase take the difference between bird and mammal as their starting point. They assume a brain able to compute nothing more than the identity transform, that is, able to store and retrieve previous inputs. On this basis, the pigeon can only respond to previously encountered exemplars. A fixed target viewed from a variety of distances and stations, however, would generate just the kind of dilated and shifted variations subsequently recognized by the pigeons in my studies. In Heinemann's model the intricacies of neural computations are replaced by the commonplaces of head bobbing and storage. My treatment of "Peanuts" (Cerella, 1980) converged on the same scheme, although in that case it was the antics of Charlie Brown rather than of the pigeon that generated the necessary exemplars.

We have then quite different models, which trade off computational complexity and storage capacity. In the first, only the prototype is stored and perspective variants are derived via computations embodied in the neuroanatomy; in the second, every variant is stored as it is encountered and only a literal match is computed. Mixed models may also be envisioned that rely partly on the prestorage of instances to cover some stimulus transformations and on the calculation of variations to cover other stimulus transformations. Given this multiplicity of models, the final appeal must be to neuroscience. In the meantime, some additional behavioral data would be relevant, as the models differ in their generative capacity, that is, in their extension to unseen instances. Whereas Heinemann points out a general correspondence between the characteristic viewing postures of the pigeon and the target deformations subsequently recognized, his case would be significantly strengthened by an experimental demonstration that restricted exposure to a target leads to a restriction in the free-viewing generalization class.

Finally, let me add my usual warning. The data presented here derive from line patterns presented to the pigeon's near-field frontal system. The findings may not apply to half-tone or even to filled patterns, much less to the far-field lateral system.

ACKNOWLEDGMENTS

Work supported by the VA Medical Research Service and National Institute of Health Grant MH32628.

REFERENCES

Cavanagh, P. (1978). Size and position invariance in the visual system. *Perception, 7,* 167–177.

Cerella, J. (1977). Absence of perspective processing in the pigeon. *Pattern Recognition, 9,* 65–68.

Cerella, J. (1980). The pigeon's analysis of pictures. *Pattern Recognition, 12,* 1–6.

Cosmides, L. (1979). *Mental rotation in pigeons.* Unpublished honors thesis, Harvard University.

Hollard, V. A., & Delius, J. D. (1982). Rotational invariance in visual pattern recognition by pigeons and humans. *Science, 218,* 804–806.

Pollen, D. A., & Ronner, S. F. (1982). Spatial computation performed by the simple and complex cells of the visual cortex of the cat. *Vision Research, 22,* 101–118.

Schwartz, E. L. (1980). Computational anatomy and functional architecture of striate cortex. *Vision Research, 20,* 645–669.

Todrin, D. C., & Blough, D. S. (1983). The discrimination of mirror-image forms by pigeons. *Perception and Psychophysics, 34,* 397–402.

Vaughan, W., & Greene, S. L. (1984). Acquisition of absolute discriminations in pigeons. In M. L. Commons, R. J. Herrnstein, & A. R. Wagner (Eds.), *Quantitative analyses of behavior: Discriminative processes* (Vol. 4). Cambridge, MA: Ballinger.

REFERENCES

[illegible]

9 Unnatural Concepts and the Theory of Concept Discrimination in Birds

Stephen E. G. Lea
Catriona M. E. Ryan
University of Exeter, UK

Much of this book is a tribute to the achievements of pigeons (and other birds) in discriminating between color slides (and other complex visual stimuli) on the basis of the objects they depict. These achievements are so surprising, at least to an imagination trained in the simple red/green or horizontal/vertical discriminations of earlier laboratory practice, that they have tended to overshadow theoretical development. They have also made it seem likely that pigeons can learn any discrimination whatsoever, given a decent training procedure.

This chapter aims to strike a slightly different note. Its major aim is to sketch the kind of theory that most researchers in the area seem to hold, if only implicitly. We show that this theory is not wholly satisfactory. To test theories of natural object discrimination, it is often helpful to look at the discrimination of less natural objects, or unnatural groupings of natural objects. Furthermore, some of these "unnatural concepts" lead to much more protracted training procedures, and in some cases birds may not be able to learn to discriminate them at all. Thus we survey some of the discriminations that pigeons either fail to learn, or learn very slowly. Perhaps this means that these kinds of learning are themselves unnatural.

A Few Matters of Definition

What, though, is meant by "natural" and "unnatural" in this context? In fact, this is only one of several distinctions whose precise meaning needs to be explored. We begin, therefore, by suggesting some ways of clarifying current terminology.

The most general way of characterizing the kinds of experiments that interest us is as *multiple stimulus discrimination.* Subjects are presented with many different stimuli and have relatively few possible kinds of response; in the limiting, and typical, case the only possibilities are to peck or not to peck. Because the number of possible responses is much less than the number of possible stimuli, the experiments can also be referred to as *category discriminations* (at least provided each different stimulus is consistently associated with reinforcement or with its absence).

We intend to avoid saying that pigeons are capable of *concept formation.* As Herrnstein (this volume) and Lea (1984) have made clear, a good deal of theoretical work is needed to sort out what concept formation actually means. On any reasonable meaning, further empirical work is needed to establish whether pigeons can do it.

On the other hand, we do not deny the existence of concepts; in fact, we accept at least the ordinary-language use of the term, and use it to define the procedure of *concept discrimination.* This describes any experimental procedure in which subjects have to distinguish between multiple stimuli, with the positive and negative sets being defined by a concept held by the experimenter. The paradigm case is, of course, Herrnstein and Loveland's (1964) experiment in which pigeons were reinforced for pecking slides that depicted people and not reinforced for pecking other slides. Herrnstein and Loveland arranged the reinforcement contingencies according to their own concept of "person."

In most concept discrimination experiments, the concepts to be discriminated have been *natural concepts,* such as person/nonperson, fish/nonfish, and so forth. In such cases, the experimenters' sorting will be derived from their history of learned language use. In a few cases, however, concepts have been generated specifically for an experiment; for example, Lea and Harrison (1978, Experiment 1) reinforced pigeons for pecking slides if at least two of three possible positive cues (green background, circular figure, black figure) were present in them. We refer to all such experiments with specially generated concepts as involving *unnatural concept discrimination.*

The simple discriminations of more conventional experimental work, in which one or two cues are necessary or sufficient for reinforcement, are limiting cases of unnatural concept discrimination. For most natural and unnatural concepts, however, there is no single objectively defined property that is either necessary or sufficient for membership of the class of instances of the concept. For this reason, Ryle (1949) referred to natural concepts as *polymorphous.* The kind of two-out-of-three rule used by Lea and Harrison (or its obvious extension to an *m*-out-of-*n* rule) is the simplest possible case of a polymorphous rule, and was studied for this reason by Dennis, Hampton, and Lea (1973). Later experimenters have tended to restrict the term *polymorphous concept* to unnatural concepts defined by an *m*-out-of-*n* rule; this restriction is probably harmless, but we try to avoid it in this chapter.

A useful control procedure is to create categories by a random sorting of stimuli. Because no rule except enumeration underlies such categories (in logical terms, they have denotation but no connotation), they are often referred to as *pseudoconcepts,* though they have obvious points in common with some unnatural concepts. It is useful to distinguish two cases. In one, a random sort is made of a set of stimuli that are all instances of some natural concept (or of no concept). We call these *random* pseudoconcepts. In the other case, it is also possible to take a set of stimuli that can readily be sorted on the basis of a natural concept and make a partition that is orthogonal to the natural one, so that each set contains both instances and noninstances of the natural concept. We call these *perverse* pseudoconcepts.

The Implicit Theory

Experimenters in the field of category discrimination have generally been sparing in their use of theory. Because of the history of theory in two closely related fields, however, two general hypotheses are always in the background. From the theory of animal discrimination learning come the ideas of absolute stimulus learning (characteristic of continuity theories in the tradition of Spence, 1936), and of dimensional learning (as used by noncontinuity theorists such as Lashley & Wade, 1946); for a summary of these theories, see Sutherland and Mackintosh (1971, Chapter 4). From pattern recognition theory comes an almost identical distinction between template matching and feature analysis (cf. Neisser, 1967, Chapter 3; Reed, 1973, Chapter 2):

The procedures used in experiments on category discrimination show an interest in eliminating the explanation in terms of absolute stimulus learning or template matching. This is the most obvious rationale for the commonly used transfer test, in which following successful training of a concept discrimination, new instances of the concept are shown to the subject. Successful discrimination of these slides shows that something more has happened than mere rote learning of the correct response to each separate instance used in training. A more sophisticated demonstration is the comparison of a concept discrimination with a perverse pseudoconcept based on a random sorting of the same stimuli. Herrnstein and de Villiers (1980, Experiment 3) carried out such a study using slides of fish and nonfish and found that the concept discrimination was learned substantially faster than the pseudoconcept. This shows that the experimenters' concept of a fish was relevant to the pigeons in some way.

The alternative theoretical account, dimensional discrimination or feature analysis, has not been subject to the same kind of implicit test. On the contrary, most experimenters probably believe that their subjects do, in some way, extract the features of the slides that make it possible to discriminate between the experimenter-defined categories. Elementary facts about the concept–discrimination literature make it possible to say a little more about the kind of feature discrimination theory that will be needed. Because the categories to be discriminated are, in most cases, defined by polymorphous concepts, birds must be able to combine the information from several features to determine the response to any given stimulus. Because many different concept discriminations can be learned, birds must be able to vary the weight they give to different features.

The simplest formalization of this theory works as follows. Several different features of each of the stimuli are measured. A weighting is assigned to each feature. Response rate is predicted from the linear sum of the weighted measure of each feature. Concept discrimination learning then consists of the adjustment of the weightings, as a result of reinforcement, until they correspond to a discriminant function that most nearly divides the stimuli into the experimenter-determined positive and negative categories.

Lea and Ryan (1983, Experiment 2) used this simple formalization of the feature-learning theory. They tested pigeons on unnatural concept discriminations, in which the stimuli were letters. They then used multiple regression to determine the extent to which response rates were related to many different features of the letters. The resulting linear models accounted for substantial proportions of the variance in response rates between the letters, both within the sets of positive and negative stimuli, and between them.

More complicated versions of the feature theory are of course possible, but until we are clearer about the empirical status of the simple version, there seems to be no justification for discussing them. In the rest of this chapter, we consider the simple version used by Lea and Ryan (1983). We refer to it as the *variably weighted linear feature model,* or, more simply, the *linear feature model.*

Advantages of the Linear Feature Model

There are good reasons why this model has implicitly guided much of the research on concept discrimination. First, as Sutherland and Mackintosh (1971) pointed out in connection with an earlier variant of it, it is consistent with much of what we know about the physiology of visual perception, about machine pattern processing, and about the early literature on animal discrimination learning. Second, it is capable of explaining both the wide variety of category discriminations that can be learned, and the speed at which some of them are learned. If the features that are useful in discriminating slides of trees from slides of nontrees, for instances, are the features that would be useful to a wild pigeon in a state of nature, then it is not surprising that pigeons learn very quickly to pick out slides of trees. Third, it can explain why some discriminations are more difficult than others. In Lea and Ryan's (1983) experiment on the discrimination of groups of letters, there were only 10 stimuli in each group (or 40 if the 4 replicas of each slide are counted). Yet learning took many thousands of trials, and at asymptote was far from perfect. Feature theory can explain this by saying that letters involve features that pigeons have not been prepared, by natural selection, to use with high priority. Similarly, the difficulty of discriminating pseudoconcepts would arise from the fact that all the naturally important features are useless for the purpose.

We suspect that it is for these kinds of reasons that experimenters have often made implicit use of feature theory. Once we start to theorize explicitly, however, feature theory has additional advantages. It is relatively easy to measure the more plausible features in pictorial displays, and to submit response rates to multiple regression analysis (or choice behavior to discriminant analysis) to establish their importance. This is the kind of procedure that we ourselves have used (e.g., Lea & Ryan, 1983).

A closer look at the evidence, however, shows that there are a number of problems with the feature theory. The rest of this chapter is devoted to considering them. To do so, we need to review some of the discriminations that pigeons do not learn very easily, or at all. Because failures to learn tend to involve long experiments that are frustrating for all concerned and are relatively hard to get published, we have to call disproportionately for them from our own unpublished work. It may be appropriate to remind readers at this stage that we can and do teach birds to discriminate concepts successfully (e.g., Lea & Ryan, 1983, Experiment 1; Morgan, Fitch, Holman, & Lea, 1976; Ryan, 1982a). In other words, the failures we describe are not to be accounted for by any general problem with the Exeter training procedures.

Those procedures are very close to those that have been described elsewhere in this book. We present slides using a random-access projector under computer control, so a new slide sequence is generated for every cycle through the stimuli. Trials consist of a 30-second stimulus presentation during which pecks at a key are recorded, but no reinforcement is given. At the end of 30 seconds, negative slides disappear immediately, but positive slides remain present until one further peck is made, whereupon grain reinforcement is given; reinforced pecks are not included in the responses analyzed. Trials are separated by an 8-second blackout, during which reinforcement is given if appropriate, and the slide projector is moved to its next position. The slide sequence is constrained so that no more than three positive or three negative slides occur in succession. Sessions typically consist of 120–480 trials. These procedures can produce accurate discrimination within as little as two sessions.

Disturbingly Easy Discriminations

We begin by considering some discriminations that are learned more easily than we might suspect from a feature theory.

One of the most striking was observed almost as an artifact by Greene (1983). For an experiment on the discrimination of repetition by pigeons, she took three successive photographs of a number of outdoor scenes. Each photograph was, at first, used consistently in a single role, either as the "initial presentation" or as the "repetition presentation" of a particular scene. Greene found that the birds were apparently discriminating repetition, but that their performance was largely based on their ability to discriminate the different photographs; randomization of the roles in which each photograph appeared completely disrupted the discrimination. Each bird saw 2 versions of each of 16 different scenes (plus 1 version of 8 others), so 16 different extremely fine discriminations were being learned.

Vaughan and Greene (1983) investigated this kind of learning in more detail. In their Experiment 1, pigeons first learned to discriminate between 2 sets of 40 slides, each randomly chosen from a set of outdoor photographs of the Cambridge, Massachusetts, area. Training was then successfully repeated with 3 further random selections without loss of performance on the original set, so that at the end of the experiment the pigeons were able to categorize 320 different slides virtually without error. This is the procedure we refer to as the "random pseudoconcept" control, because there was no "natural" categorization within the stimulus set.

However, in Vaughan and Greene's second experiment, the alternative, "perverse" pseudoconcepts were used. A random selection was made from a set of 80 slides, 40 containing fish and 40 not containing fish (78 of these were the same slides as had been used by Herrnstein & de Villiers, 1980). Within ten 160-trial sessions, the pigeons were discriminating effectively perfectly. In subsequent training, they were trained with both the orientation of the slides, and the reinforcement contingencies, reversed (so that mirror images of all positive slides were now negative, and mirror images of negative slides were positive). This new discrimination was learned within six sessions. With a little further training, the two orientations could be mixed at will without inducing errors; thus the pigeons were able to discriminate each of 80 slides from its mirror image. These results are reported in more detail by Vaughn and Greene (1984), who also report that a discrimination between 160 randomly selected slides was retained with little loss over a 2-year period in which no training or testing was given.

None of these data are logically inconsistent with a feature-learning approach. We cannot prove that there is no set of features that would allow the discriminations concerned to be made reliably. But the burden of proof must surely rest with anyone who wants to maintain that there are features that would reliably discriminate each of 16 slides from another photograph of the same scene taken seconds later with the same tripod-mounted camera, or each of 80 slides from its mirror image. Reasonably enough, Vaughan and Greene (1983) refer to their results as involving "absolute discrimination."

Disturbingly Difficult Discriminations

Next, we review some discriminations that either seem to be, or in fact were, tailor made for easy learning if the linear feature model is correct—but which prove difficult or impossible for pigeons to master.

Lea and Ryan (1983, Fig. 13–3) show the rate at which 12 pigeons acquired an artificial concept discrimination between groups of letters. There were only 10 stimuli in each of the positive and negative sets, and it is well established that letters can be discriminated by use of a relatively limited set of distinctive features (see e.g., Gibson, Gibson, Pick, & Osser, 1962; Podgorny & Garner, 1979). Yet none of the pigeons reached a fairly modest criterion of discrimination performance in fewer than 3,000 trials; 3 birds required about 20,000 trials. Figure 9.1 shows acquisition data from another, similar experiment, giving essentially the same result.

The asymptotic response rates from both these experiments were well described by linear feature models, with one important reservation. When the feature models had been fitted to the response rates, they could be used to derive predicted rates. These would inevitably be slightly different from the observed rates, but if the models were an adequate account of the discrimination, the differences should be random. In each of 12 birds in each of these experiments, rates to the positive stimuli tended to be slightly greater than predicted, and rates to the negative stimuli tended to be slightly lower than predicted. In other words, there was some residual discrimination, not accounted for by the models.

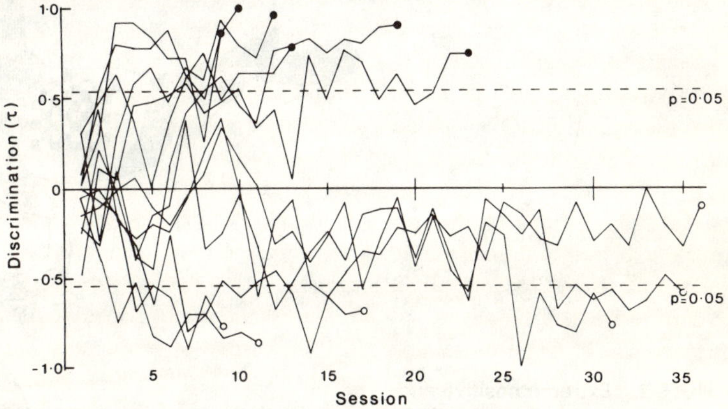

FIG. 9.1. Acquisition of an unnatural concept discrimination between groups of letters. There were 10 positive letters and 10 negative letters, and 4 copies of each were used. The vertical axis shows the correlation (Kendall's tau coefficient) between the mean response rate to a letter and its group membership; no overlap between mean response rates would give tau values of 1 or −1. Sessions consisted of 400 trials, on each of which one letter was shown. Each line shows data from one pigeon; lines ending above the horizontal axis show data for 6 birds for which letters A, B, D, E, F, K, P, Q, R, and Z were positive, lines ending below the horizontal axis show data for 6 birds for which letters C, G, J, L, N, O, T, U, V, and W were positive. These data are from an unpublished study by the authors.

The problem with these letter group experiments, of course, is that there is a wide range of features that would be of at least some use in performing the concept discrimination. We have to work back from the response rates to decide, not only which features acquired substantial control over behavior, but also which are worth considering at all. Partly for this reason, and partly because the rates of acquisition described previously seemed so slow compared with those that can be achieved with color slides of real objects, we started to work with a set of stimuli exemplified by the two creatures shown in parts a and b of Fig. 9.2. We call these *fake pigeons*. The two examples shown in Fig. 8.2 differ in five respects: wings (checkered or gray); neck (striped or gray); beak (light or dark); cere (large or small); legs (thick or thin).

In fact, we constructed a set of 64 slides, containing two copies of each of the 32 different fake pigeons that can be derived by varying these five bipolar features. Four pigeons were then trained in a 3-out-of-5 artificial polymorphous concept discrimination, in which any fake pigeon that contained at least 3 of 5 designated positive features was a positive stimulus. The beauty of this design is

FIG. 9.2. Extreme positive and negative examples of "fake pigeons" used in an unnatural concept discrimination experiment. Stimulus (a) has checkered wings, striped neck, light beak, large cere, and thick legs; stimulus (b) has gray wings, gray neck, dark beak, small cere, and thin legs. Two copies of each of the 32 possible combinations of these features were used in the experiment.

that it allows the linear feature model to be investigated systematically using a 5-way factorial analysis of variance of the response rates to the 64 slides. According to the linear feature model, all the systematic variation in the response rates should be accounted for by the main effects of the 5 features. Nonadditive combination effects of features would appear as interactions, at the first, second, or third order; response to individual fake pigeons, independent of their feature content, would appear as a fourth-order interaction; response to individual slides, independent of the fake pigeon shown on it (due, perhaps, to discrimination of dirty marks or the like) would appear as a fifth-order interaction.

Table 9.1 shows the analysis of variance for one session from one bird. Clearly, the design worked exactly as intended, and the linear feature model was fully supported. The only problem is that the bird in question was clearly only responding to one of the five features. Figure 9.3 shows that this was true of all four birds: for each of them, the mean rate of response to all stimuli containing the positive wing feature was substantially higher than the mean rate to stimuli with the negative wing feature, but none of the other features was correlated with response rate at all. In consequence, the birds' performance at the discrimination was distinctly unimpressive. The data shown in Fig. 9.3 are averages across the last 10 of 38–45 sessions, by which point performance had reached an apparent asymptote for all birds.

Of course, it could be argued that the wing feature was too striking relative to the other four. In fact, we had deliberately chosen five features of different degrees of salience. Perhaps this was a mistake on our part; perhaps pigeons cannot discriminate concepts when one relevant feature is much more salient than the others. But in this case, the linear feature explanation of other concept discriminations is suspect. If color slides of natural scenes are discriminated by integration over a number of features, these too will surely differ in salience. Our feature models of letter group discrimination (e.g., Lea & Ryan, 1983, Fig. 13–5) also assign widely different weights to different features, which is what we should expect if they have different saliences.

If pigeons really use an additive feature-learning mechanism for concept discrimination, both the letter group discrimination and the fake pigeon discrimination should be easy. The fact that they are either difficult or impossible suggests that some other or additional mechanism should be invoked. Perhaps, though, the problem with our experiments is that they use highly artificial stimuli, in comparison with the relatively naturalistic color slides that are more commonly used.

Not all pictures of natural objects are readily discriminated by pigeons, however. Ryan (1982b) attempted to train pigeons in a concept discrimination in which the positive and negative stimulus sets were photographs of two different individual pigeons (Experiment 3) or of two different individual chickens (Experiment 4). The pigeons completely failed to learn the individual-pigeon dis-

TABLE 9.1
Analysis of Variance of "Fake Pigeon" Response Rates

Source of Variation	Sum of Squares (and Variance Estimate)	Variance Ratio (F Value)
Legs	61.04	1.08
Wings	706.90	12.47
Neck	45.08	0.79
Cere	10.64	0.19
Beak	6.31	0.11
Legs.Wings	134.85	2.37
Legs.Neck	118.54	2.09
Legs.Cere	85.79	1.51
Legs.Beak	31.22	0.55
Wings.Neck	21.97	1.17
Wings.Cere	13.41	0.24
Wings.Beak	27.96	0.49
Neck.Cere	76.34	1.34
Neck.Beak	30.11	0.53
Cere.Beak	32.06	0.57
Legs.Wings.Neck	29.29	0.52
Legs.Wings.Cere	1.86	0.03
Legs.Wings.Beak	20.36	0.36
Legs.Neck.Cere	4.15	0.07
Legs.Neck.Beak	56.06	0.98
Legs.Cere.Beak	155.31	2.74
Wings.Neck.Cere	162.88	2.87
Wings.Neck.Beak	7.77	0.14
Wings.Cere.Beak	3.85	0.07
Neck.Cere.Beak	205.56	3.62
Legs.Wings.Neck.Cere	36.45	0.64
Legs.Wings.Neck.Beak	69.10	1.22
Legs.Wings.Cere.Beak	2.14	0.04
Legs.Neck.Cere.Beak	9.23	0.16
Wings.Neck.Cere.Beak	11.99	0.21
Legs.Wings.Neck.Cere.Beak	0.07	0.00
Residual (between copies)		
Sum of squares:	1760.64	
Degrees of freedom:	31	
Variance estimate:	56.79	

Note. The table shows a sample analysis of variance of mean response rates to stimuli in the "fake pigeon" concept discrimination. The data are taken from the last of 45 sessions for Bird R1. The sources of variation are the five bipolar features used to generate the stimuli and their interactions. All these sources of variation have one degree of freedom, so their variance estimates (mean squares) equal their sums of squares. The only significant source of variation is the main effect of the Wings feature.

crimination, and though some of them did learn the individual-chicken discrimination, they then failed a transfer test in which new photographs of the same chickens were used.

We still do not know why these particular discriminations were so difficult. One possibility is that the (full color) pictures of the object birds were shown to the subjects on plain white shadowless backgrounds, and Greene (1983) has shown that pigeons were strongly controlled by background cues when they were trained in a concept discrimination of an individual person. Another possibility is that the discriminations involved are just particularly difficult; certainly, human subjects require considerable training to be able to perform it (as they do to discriminate individual live chickens or pigeons). But it cannot be that the slides contained inadequate cues for discrimination, for in further experiments Ryan trained bantam cockerels to discriminate precisely the same slides of both chickens and pigeons, and they learned very quickly (Ryan, 1982a, b, Experiment 2). Clearly, pigeons do not learn all color slide discriminations easily; oddly enough,

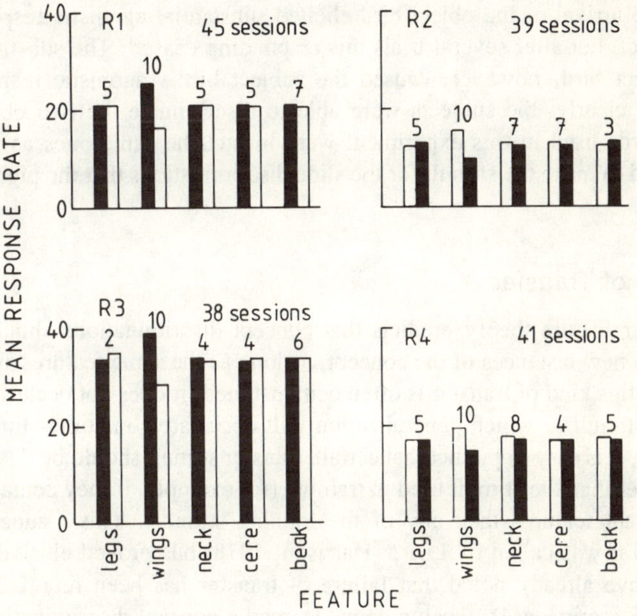

FIG. 9.3. Mean response rates to stimuli containing each value of each of the 5 features in the "fake pigeon" discrimination. Data are means of the last 10 of 38–45 sessions, shown separately for each of 4 pigeons. For each feature, the left-hand column shows response rates to stimuli containing the positive feature, and the right-hand column the rates to the negative feature. Black columns show rates to the features shown in Fig. 9.2a, whereas white columns show rates to the features shown in Fig. 9.2b. The numbers above each pair of columns show the number of sessions (out of 10) in which the mean response rate to stimuli including that particular positive feature was greater than the mean response rate to stimuli containing the corresponding negative features. Only the "wings" feature controls differential responding.

Herrnstein (personal communication) has also experienced difficulty in an experiment where a pigeon had to discriminate pictures of another individual.

It is of course possible that the social ethology of pigeons is such that they never need to discriminate individual conspecifics, and cannot in fact do so. In that case we should hardly expect them to be able to discriminate the much impoverished stimuli available in color slides. This, however, is simply not true. Anyone who has kept pigeons in an aviary knows that they behave quite differently towards different individuals within the group; most obviously, they form longlasting pair bonds. Ryan (1982b, Experiment 5) confirmed this obvious point using a dishabituation technique. Pigeons served as subjects in the experiments, and additional pigeons were used as ''object'' birds. Subject pigeons were given repeated short exposures to a single object bird. On the first few trials, the arrival of the object bird elicited substantial agonistic response from the subject, but after several trials this responding ceased. The substitution of a new object bird, however, caused the subject bird's agonistic responding to reappear; clearly, the subjects were able to discriminate the two objects. The object birds used in this experiment were in fact the same ones as were photographed to make the stimuli for the slide discriminations that the pigeons failed to learn.

Failures of Transfer

The linear feature theory predicts that concept discriminations should transfer readily to new instances of the concept, as long as the same features are present. Whereas this kind of transfer is often demonstrated, it does not occur invariably, and the stimuli to which generalization will occur are sometimes limited. Furthermore, it is easy to predict that certain transfer stimuli should be discriminated even better than the stimuli used in training (for example, if they contain more of the relevant features than any of the training stimuli). This ''superreleaser'' effect (as it was called by Lea & Harrison, 1978) has proved elusive.

We have already noted that failure of transfer has been reported by Ryan (1982b, Experiment 4), when pigeons learned a concept discrimination of individual chickens. Rather similarly, Greene (1983) trained pigeons to discriminate between slides on the basis of the number of arbitrary hand-drawn squiggles on them. The birds successfully discriminated one-squiggle slides from two-squiggle slides and generalized this apparent discrimination of numerosity to new, different, squiggles drawn by the same person as the training stimuli—but not to squiggles drawn by someone else. Presumably squiggles drawn by a different person lacked whatever features were making the original discrimination possible. These therefore cannot have included numerosity.

Similar limitation of transfer occurs in concept discriminations of other individuals. Although chickens do learn such discriminations, and do generalize to new pictures of the object bird (e.g., Ryan, 1982a), generalization fails if the transfer pictures are taken when the object birds are at very different ages from those used for the training stimuli (cf. Ryan & Lea, this volume). We have confirmed this failure of transfer in another recent experiment.

Chickens were trained in a discrimination in which the stimulus sets were pictures of other chickens taken at a series of ages, from 2 to 65 days—spanning the development in appearance from newly hatched chick to virtual adulthood. All the slides in the positive set were pictures of a single chicken, and all those in the negative set were pictures of a different chicken. The slides were back projected onto a screen at life size, and the chickens responded by pecking a transparent key through which they could see the screen.

Superficially, the chickens appeared to solve this task: Their performance was substantially better than chance. But, as the left-hand column of Fig. 9.4 shows, this was almost entirely a result of their performance on "younger" slides. The trend to discriminate only certain ages was stronger in the transfer data (on the right of the figure), which show that only at ages 12 and 23 days were the new pictures of the objects discriminated by all four subjects (the implications of this result are discussed in more detail by Ryan & Lea, this volume).

This last result is perhaps not too surprising given the feature theory, because it seems quite unlikely that there is any set of features that would make this discrimination possible. It is puzzling, however, from two other perspectives. First, there seems no reason why this discrimination should be any more difficult than the numerous pseudoconcepts discriminated by pigeons in the experiments by Vaughan and Greene (1983) and Vaughan (this volume). We have no evidence that chickens are capable of the kind of absolute discrimination described by Vaughan and Greene, but there is no obvious reason why they should be worse at it than pigeons. Second, a discrimination very like the present one is needed by chickens, and any other precocial animal with permanent social groups, in their normal life. A mother hen learns to respond selectively to her own chicks when they are very young, and maintains this discrimination through the very rapid changes of appearance that accompany their growth (Ramsay, 1951; but see some contradictory data reported by Cushing & Ramsay, 1949). Wood-Gush (1955) has reviewed this issue; we take it up experimentally elsewhere (Ryan & Lea, this volume).

Failure of the "superreleaser" transfer effect was reported by Lea and Harrison (1978). Pigeons were trained on a two-out-of-three artificial polymorphous concept discrimination, in which only the slides with exactly two positive or exactly two negative features were used in training. The "super" stimuli, with all three positive or all three negative features, were used in transfer tests. Transfer was shown, but the superpositive stimulus elicited a lower response rate than the training positives and was not preferred to them. In unpublished work in our laboratory, R. A. Joss found a similar result in a concept discrimination of letter groups. Following a feature analysis of the response rates to training stimuli, he constructed several pseudoletters that contained all the features that were correlated with high response rates; they all elicited low or zero response rates.

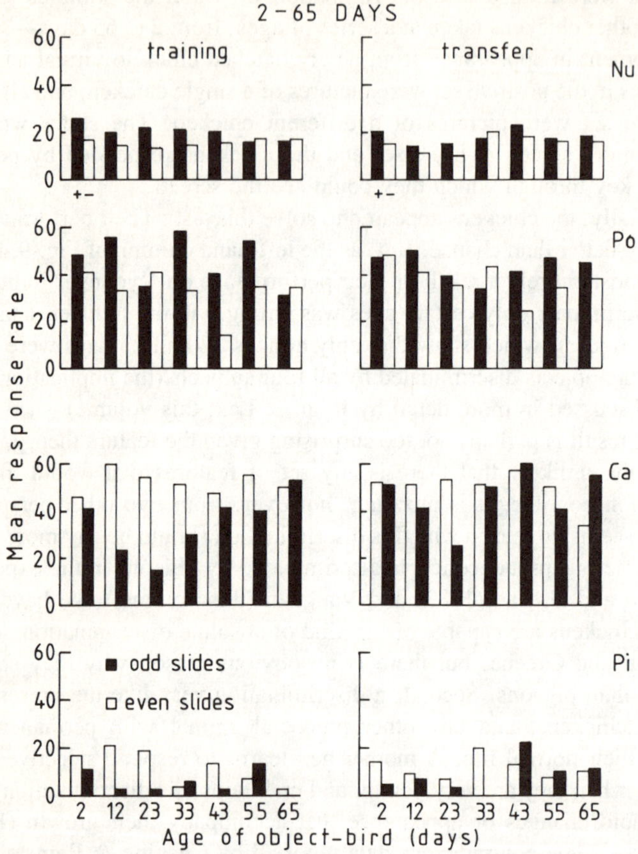

FIG. 9.4. Mean response rates of four bantam hens to training and transfer pictures of two chickens ("object birds") taken at different ages. Regardless of age, pictures of one chicken were positive, whereas pictures of the other were negative (but different object chickens were positive and negative for different subjects; the black columns in the figure always correspond to the same object bird, whereas the left-hand columns always correspond to the positive object bird). All the data are from a single session in which both familiar ("training") and novel ("transfer") pictures of the object birds were shown. The data show that the discrimination between the training stimuli extended to the pictures taken at all ages, but transfer to new pictures was not reliable.

The Features Used

If linear feature theory is correct, it should be possible to find out what features are correlated with high response rates. At an intuitive level, this can be done by inspecting the stimuli, but the process can be formalized by using multiple regression analysis, or some related statistical procedure such as discriminant analysis or analysis of variance.

The results of carrying out this sort of analysis are not encouraging for the feature theory. In the first place, the kinds of feature that emerge are not the ones that are likely to make elaborate pattern recognition possible. For example, Lea and Ryan (1983) and Lubow (1974) found that the overall luminosity of a black and white slide was a very important feature; Greene (1983) found that the background of a colored picture was more important than the supposed subject (an individual person). In our "fake pigeon" experiment, mentioned above, the feature that dominated behavior was the pattern on the wing, which was the largest and most central object; this is not the kind of property that could make possible pigeons' superb discrimination of randomly chosen slides, as is seen, for example, in the experiments of Vaughan and Greene (1983). Cerella (1980) found that pigeons' discrimination of cartoon drawings was based on the detection of extremely local features, rather than the kind of generalized or topological cue that might more plausibly be invoked to explain the recognition of polymorphous objects.

A second problem is that, if the feature theory is to work, it must allow the dominant features to vary according to the necessities of each particular discrimination. Blough (1982, 1985) carried out an experiment in which pigeons were successively taught to pick out each of the letters of the alphabet from the remaining 25; he found that a relatively small set of features could be identified that provided a good fit to all the discrimination data. The features Blough arrived at were rather different from those proposed by Lea and Ryan (1983), and he therefore argued that "we do not seem to be on the verge of a universal alphabet [of features]." But feature theory would surely require that different features would be used for identifying different letters; thus the indications at present are that the effective alphabet is more widely applicable than is comfortable for the theory.

Alternatives to Linear Feature Theory

To summarize, linear feature theory does not seem to give a good account of which discriminations will be easy or difficult for pigeons to learn; and if we nonetheless assume that it is true, it does not lead us towards a satisfactory set of effective features. But theories are not driven out by contradictory data, though given enough contradictions they come to play less and less part in our thinking. What drives out a theory is the appearance of a new and better theory. What possibilities are there?

Simple Artifact. With any new and surprising performance by a nonhuman animal, we must always be alive to the possibility of artifact. The simple control of putting blank or random slides in the projector is always worth doing, and perhaps is not always done. Nonetheless, there are now sufficient data on avian concept discrimination for us to dismiss any simple artifactual account. Many ingenious and persuasive control procedures have been devised, but the most important fact is that, in experiment after experiment, the birds' responding has been found to be lawfully related to the stimulus material within a single class of stimuli (i.e., those designated as positive, or those designated as negative, or those designated as generalization tests to be reinforced at random). It is this selective responding within a class that is our most certain guarantee that the birds are not responding to subtle cues from the apparatus, and so predicting reinforcement in ways unintended by the experimenter.

However, it is always possible that some artifact is involved in a particular experiment, especially given the very long training that is involved in some concept-discrimination experiments. Even if there is no simple artifact, such as the pigeons being able to detect photocell beams that identify positive stimuli, there is a possibility of a perceptual artifact—all positive stimuli might tend to be darker, or bluer, than all negative stimuli, for example. Theorizing should therefore, wherever possible, be based on results that have been found repeatedly, preferably in more than one laboratory, with more than one set of stimuli.

Absolute Discrimination. The second possibility is that pigeons are able to learn how to respond to large numbers of visual stimuli without needing any common process. This is the approach that is taken, for obvious reasons, by Vaughan and Greene (1983) in the light of their pigeons' capacity to learn and remember pseudoconcepts. It is plausible to suppose that an animal that has no language may be less disposed than a human being to look for or abstract common properties, and might therefore have a greater power of absolute discrimination. Before we attribute superhuman powers to the pigeon, however, it is worth recalling that people, too, can perform considerable feats of absolute discrimination; in fact we take them for granted in everyday situations. For example, if I meet one of the 200 or so students of my department in the street, I can say without effort which of the years of the course he or she belongs to; yet presumably there are no perceptual properties that are shared by all first-year students and distinguish them from second-years.

The problem with the absolute discrimination hypothesis is that it gives no account of selective generalization. Following a real concept discrimination, the subjects do generalize to new instances and not to pictures that are less similar to those that were positive in training. In fact, any demonstration that the perceptual properties of the stimulus sets make a difference is evidence against the absolute discrimination idea. If the brute force learning of instances is all that is happening, why are real concepts easier to learn than pseudoconcepts, for example? And why were our 64 fake pigeons not readily categorized?

Ecological Relevance. Perhaps what is needed is an altogether different perspective. Almost all the concept discriminations that have been both truly complex and very easily learned have involved the use of natural stimuli, in the sense of being photographs of natural objects. The surprisingly difficult discriminations, on the other hand, have mostly involved relatively abstract or artificial stimuli. Is there something about this naturalness that is of crucial importance?

One possibility is that the more natural stimuli give greater possibilities for perceptual artifacts. Greene (1983) found that pigeons that were supposedly responding to pictures of an individual person were in fact detecting cues from the background; perhaps, therefore, the reason why Ryan (1982b) could not train pigeons to recognize pictures of an individual pigeon was that she used a uniform white background (photographs of the object birds were cut out, stuck on white card, and rephotographed to make the slides).

Another possibility is that natural stimuli get some kind of preferential processing, because of their adaptive significance. Herrnstein and de Villiers (1980) tried to eliminate this possibility by using slides of fish, but perhaps a fish is in some way recognizable as a generalized vertebrate.

Of course, we have no evidence at all that birds recognize color slides as being pictures of the objects they represent (and some evidence that they do not; Trillmich, 1976). Furthermore, a photograph is a highly impoverished representation of a visual stimulus, especially given that the colors are unlikely to be properly represented for the pigeon's visual system (whose hue discriminability and color-naming functions, and hence presumably color-mixture properties, are different from those of primate vision; Wright, 1972; Wright & Cumming, 1971). But this may not matter very much if the stimuli to be discriminated contain some perceptual properties that the subjects are adapted to treat specially.

Despite the general attractiveness of the appeal to ecological relevance (cf. Kamil & Yoerg, 1982), it is not convincing in the present context. It is worth recalling Ryan's (1982b) evidence that pigeons can discriminate in life objects that they do not discriminate in slides. In a subsequent experiment, we have attempted to train pigeons in a concept discrimination in which the stimuli were two stuffed pigeons, presented in a variety of poses in a viewing chamber behind a transparent key. Four pigeons completely failed to discriminate the two dummies, or to respond to them in any way as if they were pigeons, even though they were lifelike enough to cause repeated escaped-bird alarms in the laboratory.

Prototype Theory. It is possible to compromise between feature theory and the absolute discrimination hypothesis. The result is the "prototype" theory usually attributed to Posner and Keele (1968) and Rosch and Mervis (1975) and applied in pigeons' concept discrimination by Cerella (1982). According to this idea, concept discrimination is based on the recognition of ideal, prototype forms, which are not identical to any of the stimuli but similar to all those in a particular (positive or negative) set. Individual stimuli are then categorized according to their similarity to the available prototypes. This theory is not usable until we have specified what we mean by "similarity." Rosch and Mervis suggested that it could be measured in terms of the extent to which feature content is similar to the prototype, and it is this extension that makes prototype theory a compromise between absolute discrimination and feature theory.

This theory meets some of the objections to the simple feature theory. For instance, it is plausible that it would be difficult to establish a prototype that was very similar to all the letters A, B, D, F, K, N, O, P, R, and Z, which were all members of the A-positive set in Lea and Ryan's (1983) letter group discrimination. The difficulty of creating superreleasers may be explained if what is required is for a stimulus to have similar feature values to the prototype, rather than just to have the features to a maximal extent. Nonetheless, prototype theory requires considerable modification from the form used in human cognitive psychology if it is to explain the facts of pigeon pattern learning. For example, the relative ease of pseudoconcept learning (e.g., Vaughan & Greene, 1983) is a major problem.

A Modified Prototype Theory

What form should the modification take? As a conclusion for this chapter, we argue that one simple, but unfortunately far-reaching, modification should be considered. What consequences follow if we suppose that birds can, and easily do, form and use multiple prototypes in any one situation, and in fact form as many (or as few) prototypes as are needed? This is closely similar to a proposal advanced by Cerella (1982), though he described it in terms of multiple templates rather than multiple prototypes; however, as he was prepared to allow his templates to be "flexible" and to have different "capture ratios," the difference is not important.

The resulting multiple-prototype theory is easily able to accommodate the data that have posed problems in this chapter. A pseudoconcept discrimination probably requires almost as many prototypes as there are stimuli; a real, natural concept discrimination may be performed on the basis of a small number or even one. We must also suppose that highly similar prototypes tend to coalesce—this will explain why pseudoconcepts are harder to learn than real concepts, and why one salient cue can dominate performance even though that involves a loss of accuracy (as in our fake pigeon discrimination).

Like the simple prototype theory, the multiple-prototype hypothesis needs a theory of similarity. Here, there are apparently two possibilities. One is to appeal to feature content, as in simple prototype theory; the other (taken by Cerella, 1982) is to argue that the multiple prototypes are of *local* aspects of the stimuli. Similarity between stimuli is then a matter of the proportion of prototypes they both engage. But in this version of multiple-prototype theory, the prototypes are the features, so this possibility is not distinct from the previous one. But overstressing local prototypes in fact leads multiple prototype theory to coalesce with the feature theory we have had to reject; our argument thus differs from Cerella's in that we argue for the existence of multiple *global* prototypes.

Such a multiple-prototype theory is strikingly similar to some of the artificial intelligence models of pattern recognition. For example, Carpenter and Grossberg (this volume) allow pattern categorization to involve any number of totally distinct templates but predict that it will be acquired faster if a smaller number of templates is sufficient for accurate performance.

Of course, a multiple-prototype theory is much harder to test than a simpler theory. But, as we have already noted, the study of avian concept learning has made good progress while wearing its theories rather lightly. We are not proposing that major investigations of this or any other theory should now be launched. But the experimenting community tends to proceed with a theory at the back of its collective mind, and this chapter has shown that the linear feature theory is no longer suitable for even such a cloudy role. Perhaps a multiple-prototype theory would be better.

It is, however, open to at least two important objections (several more are discussed by Cerella, 1982, though some of these apply only to the highly local templates that he considers). First, it is silent about the process by which prototypes are formed. This is true of all prototype theories. Feature models lend themselves naturally to incorporation in models of an acquisition process (e.g., Sutherland & Mackintosh, 1971); prototype models leave one either appealing to Platonic innate ideas (quite implausible for the concept of a particular individual conspecific), or helpless. The absence of a natural acquisition theory is not an objection in principle—the parallel with machine pattern recognition theories shows that ideas about prototype formation can be formulated and tested—but it is an important limitation of our present stock of ideas.

Second, the theory gives an ambiguous answer to one theoretical question that does seem important. Does avian concept discrimination involve concept formation? In cases where a single prototype is formed, a good case can be made for saying that it does, at least if we believe that the prototype has a real existence within the brain of the subject. Activation of the prototype has a one-to-one relationship to recognition of the concept. Where more than one prototype has to be invoked, however, this will not be true. We would then not want to say that the subject had formed a concept corresponding to the experimenter's, unless we had evidence of some further entity that is in some way linked to all the prototypes and is capable of entering into additional mental, or intrabrain, associations. But, as has been argued elsewhere (Lea, 1984), the existence of such a unifying entity cannot be tested by straightforward concept discriminations and transfer to new instances. Instead, it is necessary to investigate instance to category transfer in the learning of new tasks, as Herrnstein (this volume) tries to do. Such tests, however, will be greatly complicated if we have to postulate that any concept discrimination is subserved by an unspecified number of underlying prototypes, which might be anywhere between one and the number of distinct instances used in training.

ACKNOWLEDGMENTS

Much of the research on which this chapter is based was supported by the United Kingdom Science and Engineering Research Council, through project grant number GR/A82710 to SEGL, and through a research studentship to CMER.

REFERENCES

Blough, D. S. (1982). Pigeon perception of letters of the alphabet. *Science, 218,* 397–398.

Blough, D. S. (1985). Discrimination of letters and random dot patterns by pigeons and humans. *Journal of Experimental Psychology: Animal Behavior Processes, 11,* 261–280.

Cerella, J. (1980). The pigeon's analysis of pictures. *Pattern Recognition, 12,* 1–6.

Cerella, J. (1982). Mechanisms of concept formation in the pigeon. In D. J. Ingle, M. A. Goodale, & R. J. W. Mansfield (Eds.), *Analysis of visual behavior* (pp. 241–260). Cambridge, MA: MIT Press.

Cushing, J. E., & Ramsay, A. O. (1949). The non-heritable aspects of family unity in birds. *Condor, 51,* 82–87.

Dennis, I., Hampton, J. A., & Lea, S. E. G. (1973). New problem in concept formation. *Nature, 243,* 101–102.

Gibson, E. J., Gibson, J. J., Pick, A. D., & Osser, H. (1962). A developmental study of the discrimination of letter-like forms. *Journal of Comparative and Physiological Psychology, 55,* 897–906.

Greene, S. L. (1983). Feature memorization in pigeon concept performance. In M. L. Commons, R. J. Herrnstein, & A. R. Wagner (Eds.), *Quantitative analyses of behavior* (Vol. IV, pp. 209–229). Cambridge, MA: Ballinger.

Herrnstein, R. J., & de Villiers, P. A. (1980). Fish as a natural category for people and pigeons. In G. H. Bower (Ed.), *The psychology of learning and motivation* (Vol. 14, pp. 59–95). New York: Academic Press.

Herrnstein, R. J., & Loveland, D. H. (1964). Complex visual concept in the pigeon. *Science, 146,* 549–551.

Kamil, A. C., & Yoerg, S. I. (1982). Learning and foraging behavior. In P. P. G. Bateson & P. Klopfer (Eds.), *Perspectives in ethology* (Vol. 5, pp. 325–364). New York: Plenum.

Lashley, K. S., & Wade, M. (1946). The Pavlovian theory of generalization. *Psychological Review, 53,* 72–87.

Lea, S. E. G. (1984). In what sense do pigeons learn concepts? In H. S. Terrace, T. G. Bever, & H. L. Roitblat (Eds.), *Animal cognition* (pp. 263–276). Hillsdale, NJ: Lawrence Erlbaum Associates.

Lea, S. E. G., & Harrison, S. N. (1978). Discrimination of polymorphous stimulus sets by pigeons. *Quarterly Journal of Experimental Psychology, 30,* 521–537.

Lea, S. E. G., & Ryan, C. M. E. (1983). Feature analysis of pigeons' acquisition of discrimination between letters. In M. L. Commons, R. J. Herrnstein, & A. R. Wagner (Eds.), *Quantitative analyses of behavior* (Vol. IV, pp. 239–253). Cambridge, MA: Ballinger.

Lubow, R. E. (1974). High-order concept formation in the pigeon. *Journal of the Experimental Analysis of Behavior, 21,* 475–483.

Morgan, M. J., Fitch, M. D., Holman, J. G., & Lea, S. E. G. (1976). Pigeons learn the concept of an "A." *Perception, 5,* 57–66.

Neisser, U. (1967). *Cognitive psychology.* New York: Appleton-Century-Crofts.

Podgorny, P., & Garner, W. R. (1979). Reaction time as a measure of inter and intraobject similarity: Letters of the alphabet. *Perception and Psychophysics, 26,* 37–52.

Posner, M. I., & Keele, S. W. (1968). On the genesis of abstract ideas. *Journal of Experimental Psychology, 77,* 353–363.

Ramsay, A. O. (1951). Familial recognition in domestic birds. *Auk, 68,* 1–17.

Reed, S. K. (1973). *Psychological processes in pattern recognition.* New York: Academic Press.

Rosch, E., & Mervis, C. B. (1975). Family resemblances: Studies in the internal structure of categories. *Cognitive Psychology, 7,* 573–605.

Ryan, C. M. E. (1982a). Concept formation and individual recognition in the domestic chicken (Gallus gallus). *Behavior Analysis Letters, 2,* 213–220.

Ryan, C. M. E. (1982b). *Mechanisms of individual recognition in birds.* Unpublished master's thesis, University of Exeter.

Ryle, G. (1949). *The concept of mind.* London: Hutchinson.

Spence, K. W. (1936). The nature of discrimination learning in animals. *Psychological Review, 43,* 427–449.

Sutherland, N. S., & Mackintosh, N. J. (1971). *Mechanisms of animal discrimination learning.* New York: Academic Press.

Trillmich, F. (1976). Learning experiments on individual recognition in budgerigars (Melopsittacus undulatus). *Zeitschrift für Tierpsychologie, 41,* 372–395.

Vaughan, W., & Greene, S. L. (1983). Acquisition of absolute discriminations in pigeons. In M. L. Commons, R. J. Herrnstein, & A. R. Wagner (Eds.), *Quantitative analyses of behavior* (Vol. IV, pp. 231–238). Cambridge, MA: Ballinger.

Vaughan, W., & Greene, S. L. (1984). Pigeon visual memory capacity. *Journal of Experimental Psychology: Animal Behavior Processes, 10,* 256–271.

Wood-Gush, D. (1955). The behaviour of the domestic chicken: A review of the literature. *British Journal of Animal Behaviour, 3,* 81–110.

Wright, A. A. (1972). Psychometric and psychophysical hue discrimination functions for the pigeon. *Vision Research, 12,* 1447–1464.

Wright, A. A., & Cumming, W. (1971). Color-naming functions for the pigeon. *Journal of the Experimental Analysis of Behavior, 15,* 7–17.

10 Visual Categorization by Pigeons

Robert G. Cook
Anthony A. Wright
University of Texas Health Science Center, Houston

Donald F. Kendrick
Middle Tennessee State University

INTRODUCTION

Animals frequently display adaptive behaviors in response to novel stimuli. Often these behaviors have been previously exhibited in response to similar, but clearly different, stimuli and contexts. This grouping of discriminable stimuli into the same behavioral category is the primitive essence of conceptual and inductive processing. This chapter examines the categorization and representation of two real-world categories by one simple processing system, the pigeon. Recently pigeons have demonstrated their capacity to categorize a wide variety of stimuli, for example, oak leaves, fish, trees, water, people, and alphabetic characters (Cerella, 1979; Herrnstein & de Villiers, 1980; Herrnstein, Loveland & Cable, 1976; Lea & Harrison, 1978). A different facet of stimulus classification by pigeons is examined in each of the three sections of this chapter. In the first section the influence of the number of training stimuli and their correlation with reinforcement on the acquisition and transfer of a categorical discrimination are investigated. In the second section, the issue of stimulus and exemplar equivalence within a category is examined. An analysis of the stimulus features that control the discrimination of the two categories is undertaken in the third section of the chapter.

One distinctive aspect of this research is the use of line drawings depicting birds and mammals as the discriminated categories. Examples of the animal figures used in these experiments are given in Fig. 10.1. These drawings were collected from scientific journals and magazines of the 19th century (Harter, 1979; Iyari, 1979).

There are several advantages in using these particular stimuli. Perhaps, the

FIG. 10.1. Twelve examples of each of the 2 categories used in the experiments described in this chapter. The top 3 rows show examples of the bird category, which decrease in their judged typicality from left to right and from top to bottom (see text for details). The birds judged to be most and least prototypical are displayed in the upper left and lower right, respectively. The bottom 3 rows show examples of mammals according to their judged typicality. The mammals judged to be most and least prototypical are displayed in the upper left and lower right, respectively.

most important is the fact that the bird and mammal categories represent naturally occurring categories, at least to humans. The drawings are accurate, detailed, and rich in information about the animals they represent. Because of their educational purpose, they often show the animals in characteristic postures and habitats, making it possible to extract information about each category over and above the simple form of each animal. Differences in the orientation of the animals, the artist's perspective, the number of animals drawn, and the subject–viewer distance add irrelevant variation to the stimuli.

The form of these stimuli further allows precise experimental manipulation of their features. Because each drawing is a collection of pen strokes, each line can be individually altered when necessary. The use of black and white line drawings also excludes color information, which can overshadow the processing by pigeons of other attributes in complex stimuli (Lamb & Riley, 1981). These stimuli are unusual because they allow experimental inquiry with naturally occurring complex stimuli that are simultaneously easily manipulated and controlled by the experimenter. We believe the choice of these stimuli is an important contribution to the findings presented in this chapter.

The basic organization of the discrimination used in these experiments is similar to those previously used in investigating categorization in pigeons. The pigeons discriminate between two sets of stimuli in a classical conditioning variation of the go/no-go discrimination task. The S+ stimuli predict food reinforcement, whereas the S− stimuli precede extinction or a timeout. Pigeons rapidly learn to peck only the stimuli associated with food reinforcement and do not peck the S− stimuli. Following acquisition of the discrimination, transfer tests with novel examples of each stimulus class are conducted. Pecking behavior in response to the novel examples is used to determine the presence or absence of a categorical discrimination.

VARIABLES CONTROLLING ACQUISITION AND TRANSFER

Categorizing and grouping similar stimuli represents an excellent way for simple neural systems to compress the amount of information to be retained. This first section of the chapter examines two factors that may influence the process by which a categorical discrimination is learned. One factor is the degree of correlation between category membership and reinforcement. The other is the number of exemplars used to typify a category during training. In conjunction, these variables can help us determine the nature of the categorical representation used by the pigeons.

For example, one way to represent a category is to extract only the salient common features and then act in the same way to novel stimuli having these features. Such a mechanism might be revealed through evidence of faster acquisition of a categorical discrimination when reinforcement is correlated with category membership and a particular set of features, than when it is not (Herrnstein & de Villiers, 1980). Categorical discriminations of this type may consist of two phases. The first, or "stimulus learning," phase would consist of assessing specific stimulus configurations and their associated reinforcement characteristics, essentially the gathering and storage of basic information about the exemplars. A second, or "concept" phase, in which relational information about common properties is extracted, could then takeover. It is during this concept phase that relational information between the exemplars could begin to reinforce and feed back on each other, resulting in a facilitation of acquisition.

The facilitation produced by such a concept phase can be contrasted with representational processes in which no relational information about the exemplars is extracted. Such a process essentially restricts learning to the "stimulus" phase, with each exemplar being learned about separately and distinctly. In this case, transfer to novel stimuli would be mediated by generalization or item similarity to the previously learned exemplars. This exemplar-based mechanism would be particularly effective for categories having highly similar members (Rosch, Mervis, Gray, Johnson, & Boyes-Braem, 1976). Such stimulus-specific exemplar learning would be required of any condition where no concept unifies the S+ or S− stimulus sets. One way to produce such a condition is to use arbitrary sets of stimuli and randomly designate their reinforcement value (Vaughan & Greene, 1983a). In this way, reinforcement cuts across natural category boundaries, making relational or conceptual information of no value. Acquisition would therefore have to be limited to learning specific stimulus configurations. Such extensive stimulus-specific learning is clearly not beyond the pigeon, which can be taught to discriminate over 400 arbitrary stimuli (Vaughan & Greene, 1983b).

Experiments examining the relation between category acquisition and the correlation of reinforcement with category membership have produced mixed results. Herrnstein and de Villiers (1980) found that acquisition of a discrimination using fish and nonfish stimuli was achieved faster when reinforcement was correlated with category membership than when it was not, suggesting that the discrimination consisted of more than learning the specific instances. Similar results have recently been found by Edwards and Honig (1987), and Wasserman, Kiedinger, and Bhatt (1988). On the other hand, Vaughan and Greene (1983a) examined a similar comparison between experiments, which suggested no facilitation in acquisition rate with correlated reinforcement. Schrier, Angarella, and Povar (1984) also found no differences in acquisition rate between correlated and uncorrelated procedures in tests conducted with monkeys. These latter two results suggest that the animals were learning only about the specific stimuli in the discriminations.

An important factor in determining when correlated facilitation will occur may be the number of exemplars used to represent the category during training. Studies of human concept development have shown that the use of many exemplars during training enhances concept learning and transfer. A small number of training exemplars, on the other hand, appears to promote the learning of individual stimuli (Homa, Sterling, & Treple, 1981; Omohundro, 1981). Extrapolating these human data to the animal experiments, the use of only a few examples might dispose the animals to learn limited and specific features about the stimuli, whereas encountering a larger set of items during training may encourage the development of a more relationally based representation of the category.

The following experiment examines both these issues, namely, the effects of correlated reinforcement, and the training set size. Groups of pigeons were trained with either 5 or 35 stimuli from the bird and mammal categories. These two groups were further divided according to whether or not reinforcement was correlated with the categories. Thus, for half the pigeons, reinforcement was correlated with naturally occurring category membership, whereas for the other half, reinforcement was uncorrelated with category membership. Category development could then be assessed by both acquisition rate and degree of transfer in the following four groups of five White Carneaux pigeons: *5C*–5 exemplars of each category correlated with reinforcement, *35C*–35 exemplars of each category correlated with reinforcement, *5U*–5 exemplars of each category randomly assigned with respect to reinforcement, and *35U*–35 exemplars of each category randomly assigned with respect to reinforcement.

The stimuli used were slides reproduced from the edited collection of Iyari (1979), which were rear projected onto a centrally located 2.5 cm × 2.5 cm response key of a standard operant chamber. A daily session consisted of 140 discriminative classical conditioning trials, with equal numbers of S+ and S− trials in each session. Each trial consisted of a 6 s presentation of an item followed by food reinforcement on S+ trials, and no reinforcement on S− trials. A variable intertrial interval separated the trials. For the 2 correlated groups (5C and 35C), delivery of reinforcement followed presentations of bird pictures (S+), but not mammals (S−). For the uncorrelated groups, each bird and mammal picture was randomly assigned to be either followed by reinforcement or not (5U: 3 birds and 2 mammals were S+; 35U: 18 birds and 17 mammals were S+; the remaining stimuli in each group were S−). The pigeons received daily training sessions until the first session in which they directed 70% of their total responses to the S+ stimuli. At this point the pigeons were tested for category transfer with 40 novel examples of birds and mammals. Ten transfer trials (5 from each category) were conducted in each of 4 consecutive sessions.

Figure 10.2 presents the rates of acquisition for the 4 groups as a function of sessions. The 5C and 5U groups both learned more quickly than either 35C or 35U groups. There were no reliable differences between the uncorrelated and correlated groups ($F(1,16) = 0, p > .05$). Thus, the correspondence of reinforcement with category membership did not produce faster acquisition of the discrimination.

Table 10.1 presents the results from the transfer tests. These transfer tests

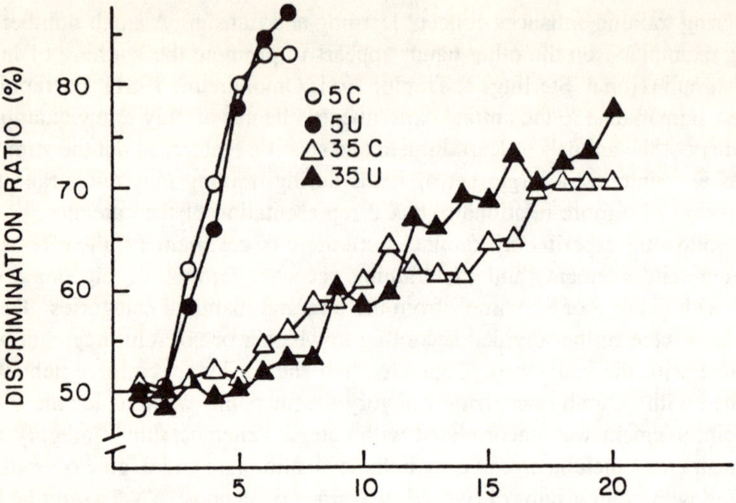

FIG. 10.2. Acquisition curves for 4 groups of pigeons. 5C: Group trained with 5 exemplars of each category and reinforcement correlated with category membership. 5U: Group trained with 5 exemplars of each category and reinforcement not correlated with category membership. 35C: Group trained with 35 exemplars of each category and reinforcement correlated with category membership. 35U: Group trained with 35 exemplars of each category and reinforcement not correlated with category membership.

TABLE 10.1
Effects of Training Set Size and Correlation Between Category Membership and Reinforcement on Acquisition and Transfer of Bird/Mammal Discrimination

Group	# of Stimuli	Type of Training	Sessions to Criterion	Transfer Peck Rate Bird	Mammal
5U	5	Uncorrelated	5.2	13.8	11.3
5C	5	Correlated	4.4	10.7	10.3
35U	35	Uncorrelated	14.6	14.2	13.9
35C	35	Correlated	15.4	17.3	13.6

found that the two uncorrelated groups failed to discriminate between novel exemplars of each category. Among the correlated groups, only the group with the larger number of training exemplars (35C) showed transfer, with significantly lower peck rates to novel mammals relative to novel birds ($t(8) = 5.9$, $p <$ .05).

These results begin to both clarify and narrow the range of mechanisms potentially responsible for stimulus classification in pigeons. Three general classes of mechanisms have been proposed as underlying categorical behavior

(Lea, 1984; Medin & Smith, 1984). The *classical* view states that there is a specific set of necessary and sufficient conditions defining a class. Simple inclusion of a stimulus within a category is determined by the presence and absence of defining features. *Prototype* and *parametric* models suggest that a category is represented by the features or properties that are highly associated within its members. Those features that have the highest validity or occurrence are derived, and in some theories transformed into an abstract representation or standard for the category (Mervis & Rosch, 1981). Thus, in contrast to the classical models, this set of models suggests that the boundaries of categories are fuzzy and indistinct, because some attributes or features may be found in more than one category. *Exemplar* or *instance* views are characterized neither by a set of defining features nor by a set of differentially weighted features, but by individually retained exemplars. In these models, categories are represented by a series of specific stimulus configurations, against which any new exemplars are compared (Estes, 1986).

The results from the acquisition experiment described previously seem most consistent with an instance or exemplar view as the classification process used by our pigeons. The identical rates of acquisition for the uncorrelated and correlated groups, when controlling for the number of training stimuli, suggests that the pigeons required a certain number of presentations to learn a set of particular configurations, regardless of whether a concept unified the stimuli or not. Increasing the number of training stimuli from 5 to 35 did not appear to facilitate acquisition or modulate the occurrence of facilitation due to the correlation of reinforcement with category membership. The control of acquisition rate by the number of item presentations and the lack of correlated facilitation both are consistent with the idea that the pigeons were learning only specific stimulus configurations in all four conditions. That some pigeons profited from a larger training set, as indicated by the significant transfer in the 35C or many example group, is most likely due to the increased probability of a training exemplar being similar to a transfer item.

Given an instance or exemplar interpretation, why are pigeons capable of generalizing from training exemplars to transfer items, but not between the training exemplars themselves? One consideration is the apparently higher weighting given absolute factors over relational factors in discriminations by pigeons (Carter & Werner, 1978; Premack, 1978). The processing of absolute stimulus factors may be encouraged by the use of a small number of stimuli that rapidly become highly familiar. Because of their extensive experience with the training exemplars, the pigeons may come to process these items as very distinct from one another. It is only when unfamiliar stimuli are tested do pigeons begin comparing new exemplars with their repertoire of stored exemplars.

An additional factor that may also have encouraged the learning of absolute stimulus configurations is the use of an unfamiliar environment, namely the operant chamber. A quite reasonable strategic priority in a new situation is simply to describe and inventory the environment. Given no information about the extent and distribution of possible stimuli and their relations in an operant chamber, a reliance on an absolute or nonanalytic strategy by the pigeons might not be unexpected (Flannagan, Fried, & Holyoak, 1986). Perhaps this experiment has captured this initial reliance on absolute and stimulus-specific factors, and more relational or analytic processes would become more important only through continued experience.

STIMULUS NONEQUIVALENCE AND CATEGORY MEMBERSHIP

Classical views of concept formation suggest that categories are determined by the subjects according to a set of necessary and sufficient criteria (Medin & Smith, 1984). Implicit in this view is the idea that any exemplar should represent the category as well as any other, making the category boundaries clear and distinct. In contrast, research on categorization by humans has found that different exemplars are not equally representative of a category. Humans reliably rate some examples as better or more prototypical than other examples of the same category (Rips, Shoben, & Smith, 1973; Rosch, 1975). For instance, humans consistently judge a robin as a better or more typical example of a bird than a penguin. Such results have suggested that a gradient of ''typicality'' exists in both perceptual and semantic categories. Typicality gradients have been found to affect human reaction time, similarity ratings, error rates, and priming effects, making this ubiquitous-graded structure a hallmark of categorization by humans (Rosch, 1978). The question as to whether animals also have a graded organization within their visual categories, then, becomes a natural one to ask.

Herrnstein and de Villiers (1980) rank ordered pigeon and human discriminations of pictures of fish from nonfish to investigate the degree to which humans and pigeons similarly perceived the categories. They found no reliable correlation between the ordering of the pigeons (by peck rate) and those of the humans (by reaction time). Significant correlations existed within each species, suggesting that each was attending to different properties of the stimuli. Herrnstein and de Villiers suggested that for the humans these features appeared to be the figure/ground contrast, prototypicality, and canonical nature of any particular slide. For the pigeons only the first and third factors seemed important, with an additional difficulty with slides containing only partial views of fish.

Schrier, Angarella, and Povar (1984) examined concept formation in stumptailed monkeys using a category based on human figures. They found that slides portraying good examples of humans were better classified by the monkeys than those containing poor examples. Their good slides contained ''relatively close, full figures or head and shoulder of humans viewed frontally,'' whereas in the poor slides "the humans were small, often in the periphery, or partially obscured." Although neither of these studies directly examined the issue of graded category membership, they demonstrate that stimulus qualities beyond mere category membership influence classification.

In the next series of experiments we investigated the issue of stimulus equivalence and exemplar typicality in pigeons. The design of the experiments involved three steps. The first step consisted of having human subjects rank in order sets of bird and mammal line drawings according to their prototypicality. These ratings were then used as a measure of each instance's degree of typicality and as a baseline for judging the responses of the pigeons. The second step involved training the pigeons using stimuli from a selected portion of the human-derived typicality gradient. The third step consisted of transfer tests, using stimuli from along the entire human typicality gradient. If the pigeons are sensitive to the same factors that determined the human typicality rankings, then transfer performance should be directly related to the human typicality gradient. This would be reflected in a graded pattern of transfer to novel exemplars, depending on where the test item was selected from the typicality continuum. If pigeons are not sensitive to the same factors, or use cues entirely different than humans, then transfer performance should be unrelated to the human typicality rankings.

Eight human subjects rank ordered 60 examples of the bird and mammal category based on the instruction of "rank-order the following category from most representative to least representative, from the best examples to the worst examples, from the most prototypical to least prototypical." The items in Fig. 10.1 are arranged according to these ratings, with items becoming judged less typical as you move from left to right and then downward in each row. For the bird category, the average rank order correlation among the subjects was $r = .82$. The results closely agree with the semantic typicality ratings reported by Uyeda and Mandler (1980; $r = .83$ for the 13 common exemplars that could be identified between their experiment and the present one).

For the mammal category the results were different. The mean rank-order correlation among the eight subjects was only $r = .06$. This low average correlation was the result of two opposing clusters of four subjects each, with group averages of $r = .53$ and $r = .68$. The source of the difference is illustrated by the ranking of the single primate exemplar (the group of monkeys in Fig. 10.1). One set of four subjects saw this picture as a good example of the mammal category, whereas the other group consistently ranked it as a poor exemplar. All subjects complained of the difficulty in ranking the mammals, perhaps because mammals represent a heterogeneous and superordinate category for humans (Rosch et al., 1976). Because the combined rankings for all eight subjects were highly correlated with the previous semantic typicality judgments reported by Uyeda and Mandler (1980; the rank-order correlation of the 15 common exemplars between the two experiments was $r = .70$), we decided to use the overall judgments in these experiments.

The rank-ordered exemplars were then divided into three 20-item groups. The top 20 items were designated as good examples, the bottom 20 as poor examples, and the middle 20 as intermediate examples. Two groups of three pigeons were then trained to discriminate birds from mammals using either the good or poor examples of the bird category. The contrast category for both groups consisted of the intermediate level of the mammal category. After acquisition, the pigeons were tested for transfer with the exemplars not used in training.

The six pigeons were trained in an operant chamber using a discriminative classical conditioning procedure. The front panel contained a large clear Plexiglas window (9 cm high × 9.3 cm wide) that served as the response key. Slides were rear projected onto a screen (12 cm high × 18 cm wide) located 25 cm behind the large response key. A trial consisted of an 8 s stimulus presentation followed automatically by the reinforcement contingencies. Slides of birds were followed by a 2 s access to mixed grain, whereas slides of mammals were followed by a 15 s dark timeout. Trials were separated by a variable 15 to 21 s intertrial interval. A daily session consisted of 80 trials with each of the 40 training stimuli (20 birds and 20 mammals) presented twice. Transfer tests were conducted when the S+ trials (birds) received 80% or more of the responses for five consecutive sessions. During the transfer trials, novel examples of birds and mammals were included in the second half of a session. A total of 32 bird and 38 mammal transfer examples were tested in two blocks of five sessions each, as some exemplars from each category were excluded because of their high similarity to specific training exemplars.

Baseline trials during the transfer sessions showed significantly higher peck rates to pictures of birds than to pictures of mammals ($F(1,20) = 30.9$, $p < .001$). Transfer trials showed higher peck rates to novel bird trials than to novel mammal trials ($F = 5.048$, $p < .036$). Because there were no systematic effects from retesting the items ($Fs(1,20) > .26$, $p > .61$), the data from both presentations were used. On both baseline and transfer trials, bird pictures were pecked at earlier and more frequently than mammal pictures. Presented in Fig. 10.3 are the baseline and transfer results for three response measures: average response rate, the percentage of trials receiving at least one peck, and latency to first peck.

Our main interest was in determining if transfer performance in pigeons was correlated with the human typicality rankings. If the pigeons transferred equally well to all examples, it would indicate equivalence between the exemplars and the absence of any graded structure within the categorical representation of the pigeons, at least when compared with humans. On the other hand, if transfer was influenced differentially by an item's position on the typicality gradient, it would indicate that pigeons and humans share a common perception of these categories.

The results for the bird category are presented in Fig. 10.4. The peck rate, percentage of trials pecked, and response latency as a function of those typicality classes not used during training are given. The basic pattern of results can be understood by examining the top panel. The pigeons trained with good examples

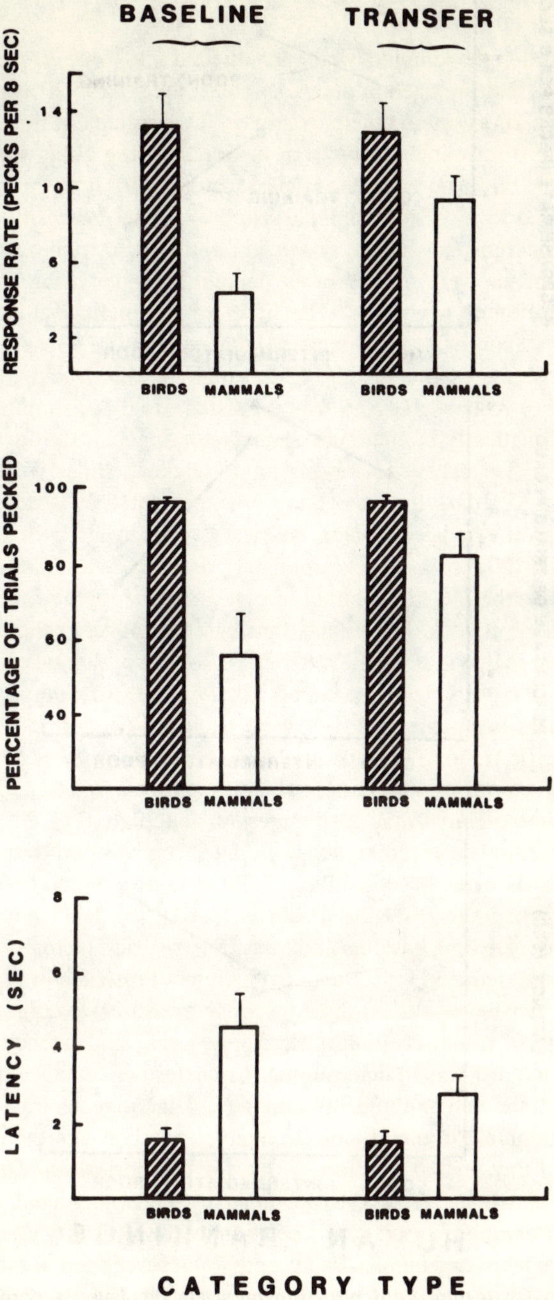

FIG. 10.3. Results from three response measures for baseline and transfer trials during first typicality experiment. Results from the two groups of pigeons were combined.

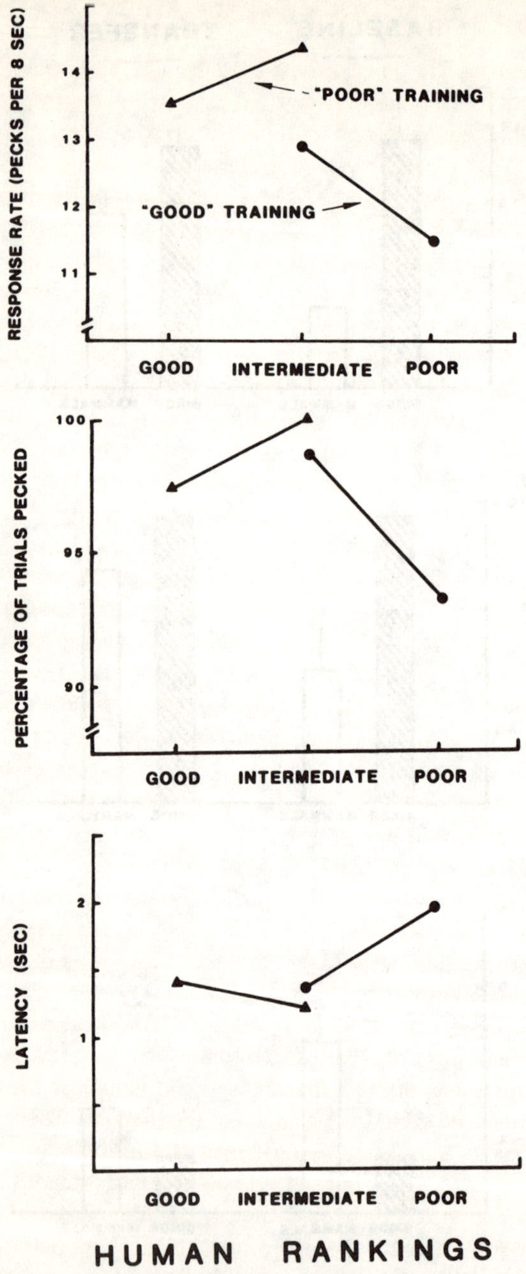

FIG. 10.4. Results from bird transfer trials displayed according to their typicality, as judged by humans. Diamonds represent data collected from pigeons trained with "poor" exemplars, circles represent data from pigeons trained with "good" exemplars. With regards to latency to first peck, note that good stimulus control is indicated by a pattern of data the inverse of that seen with peck rate.

of birds pecked more at the intermediate instances of birds than the poor instances. This pattern is what might have been expected given the human ratings of the category. The pigeons trained with poor examples reveal quite a different pattern, however. This group also pecked more frequently at the intermediate instances, but less to good instances. The other response measures show the same pattern of results for both groups, easily summarized by the observation that each group pecked most frequently and with a shorter latency to those exemplars of the bird category ranked nearest their respective training exemplars ($Fs(1,5)$; peck rate = 3.9, $p < .09$; latency = 7.9, $p < .036$; percentage of trials pecked = 7.3, $p < .042$). Ranking the transfer exemplars by peck rate and comparing them with the human rankings revealed a positive rank-order correlation ($r = .29$) for the pigeons trained with good exemplars and a negative correlation ($r = -.30$) for pigeons trained with poor exemplars. These correlations are comparable to those obtained between the pigeons themselves, in each group (ranging from .27 to .35). This correlation between degree of transfer and the human typicality ratings is noteworthy in light of previous negative findings. For instance, Herrnstein and de Villiers (1980) found an average correlation of only $r = .04$ in their similar analyses of the fish/nonfish discrimination.

Figure 10.5 presents the transfer results for the mammal category. As in the bird category, there was again differential transfer as a function of the human typicality ratings. As was mentioned earlier, mammals were the S− stimuli in these experiments, meaning that longer latencies and low peck rates are indicative of good stimulus control for this class. Good instances of the mammals resulted in better control than the poor instances: lower peck rates ($F(1,5) = 16.3, p < .011$), fewer trials being pecked ($F(1,5) = 8.2, p < .034$), and longer response latencies ($F(1,5) = 15.6, p < .011$).

The differential transfer for both the bird and mammal categories suggests that stimulus information about each category controlled the discrimination; that is, peck rate was not determined by a single category and a default response to all other stimuli. Of greater importance, the transfer performance of the pigeons was correlated with the typicality ratings of humans. This indicates the presence of properties within the categories that are used by both species to classify the stimuli. Also of interest is the fact that the specific nature of the stimuli used during training influenced the transfer results of the pigeons. Those trained with only "poor" exemplars transferred best to lower ranked items, whereas pigeons trained with "good" exemplars transferred best to highly ranked items.

The use of an entire 20-item subset during training unfortunately left no instances available for testing that portion of the typicality spectrum. To examine exemplars from within the scope of training stimuli, and to examine further more points on the typicality gradient, a second experiment was conducted. Additional stimuli were selected from another sourcebook of animal line drawings (Harter, 1979), adding 193 new birds and 248 new mammals to the original set. Unlike in the first set, some species were represented in more than one drawing.

Nine human subjects, four of whom had rated the first set, ranked the new

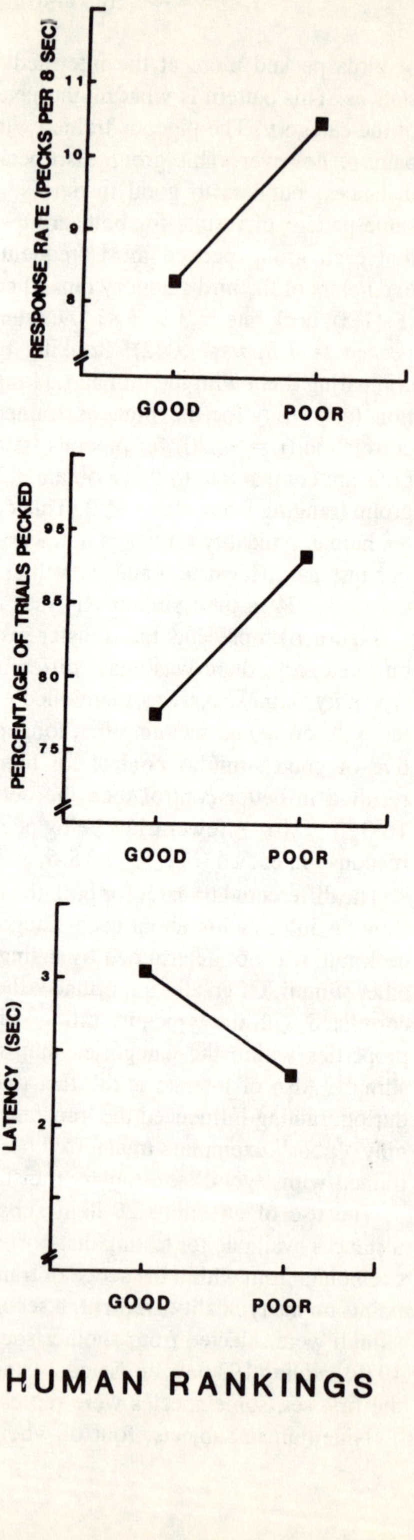

FIG. 10.5. Results from mammal transfer trials displayed according to their typicality, as judged by humans. The results from the two groups of pigeons were combined, because both groups were trained with the same set of mammal exemplars. With regard to latency to first peck, note that good stimulus control is indicated by a pattern of data the inverse of that seen with peck rate.

examples along with the original set. The correlation between subjects on the bird rankings was not as high as in the first experiment (mean $r = .57$). The increased pool size, however, facilitated agreement within the mammal category relative to the first experiment (mean $r = .37$). The addition of more examples appeared to stabilize the mammal rankings. Recall that in the first experiment the primate example was ranked near the top by four of the subjects and near the bottom by the remaining four subjects. Additional examples of other primates in the second experiment prevented this disparity, evidently by narrowing the range of different categorical structures that the human subjects could adopt.

The second transfer experiment was conducted approximately 3 months after the first. During the intervening period, one pigeon in the "good" bird training group died for reasons unrelated to the experiment. The remaining pigeons were tested for transfer with 68 new instances from each category over the course of 15 sessions. These transfer tests revealed a pattern of performance similar to that seen in the first experiment. Figure 10.6 shows mean peck rates for transfer items for the pigeons trained with good bird exemplars, those trained with poor bird exemplars, for both the bird and mammal categories as a function of the human typicality rankings. In general, stimulus control was best for the exemplars ranked closest to those used in training. Correspondingly, the poorest stimulus control was for the exemplars ranked furthest from those used in training.

The top panel show the results for the pigeons trained with good bird exemplars and basically resemble the findings of the first experiment. The middle panel shows performance for the pigeons trained with poor bird exemplars. These pigeons also transferred best to exemplars ranked close to their training examples, with a downward slope toward the good exemplars, as in the first experiment. The decrease in stimulus control to either side of the mean training rank should be noted. The second set of stimuli contained several new examples of birds judged to be even poorer than those of the first set. The original training stimuli were still used, however, for baseline training prior to and during transfer testing. Thus, it is of interest to observe the loss in stimulus control by these "very poor" stimuli also. A similar U-shaped function is also seen in the transfer results for the mammal category, which are presented in the lower panel. Again, stimulus control decreased as the typicality distance from the mean ranking of original training items increased. As in the case of the "poor" birds, training with intermediate values produced a U-shaped function of stimulus control.

These experiments generated two important new results. The first of these is the presence of a structure in both categories to which the humans and pigeons were sensitive. The sensitivity of both species to this stimulus structure produced a correspondence between the transfer performance of the pigeons and the typicality gradients of the humans. Second, the transfer results of the pigeons were influenced by the type, range, and typicality of the stimulus items used in training; that is, pigeons trained with "poor" exemplars transferred best to poor exemplars and pigeons trained with "good" exemplars transferred best to good

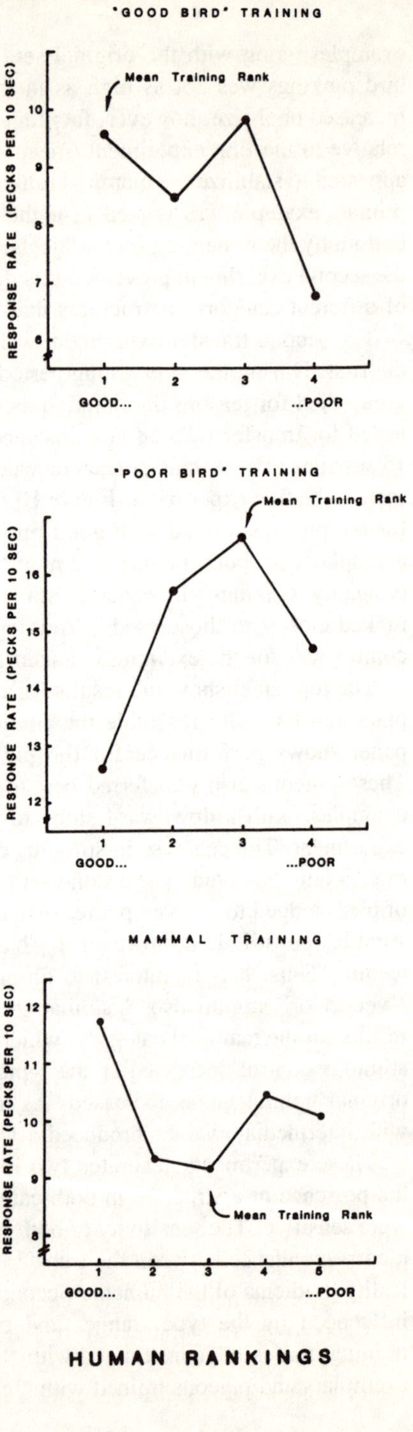

FIG. 10.6. Results of bird and mammal transfer trials as a function of their ranking by humans for their typicality of the category. The points along the human typicality gradient represent approximately equal divisions of the total number of stimuli. The mean training rank was determined by averaging the ranks of the 20 stimuli used during baseline training. The top panel depicts bird category transfer for pigeons trained with "good" exemplars. The middle panel depicts bird category transfer for the pigeons trained with "poor" exemplars of birds. The bottom panel depicts mammals category transfer for both groups of pigeons.

exemplars. Both groups transferred best to mammals judged to be intermediate in rank and closest to the training exemplars.

Typicality gradients have been interpreted as evidence against definitional models of categorization by humans (Rosch, 1978; but see Bourne, 1984). By simple extension, the graded transfer of the pigeons would suggest that their representations of the categories are also not based on a set of defining features or exact boundary conditions. The results are more consistent with representational systems based on parametric feature analysis or the retention of specific exemplars.

Several explanations have been proposed for the typicality effect in categorization by humans. The primary ones are differences in family resemblance among exemplars, subject familiarity with the exemplars, or knowledge about the natural frequency of the exemplars (Ashcraft, 1978; Barsalou, 1985; Malt & Smith, 1982; Rosch & Mervis, 1975). Item familiarity and/or knowledge about their natural frequency seem unlikely explanations for these results, because these pigeons have no experience with the depicted animals. Because of this consideration, explanations based on information from the stimuli themselves would seem required.

Rosch and Mervis (1975) conducted a series of experiments with humans examining the relation between typicality and the family resemblance of the category items. Exemplars that shared the most attributes or elements with all other members of the category and the least overlap with contrasting categories were judged to be the most prototypical examples of the category. Thus, for humans, the good or prototypical items of a category have properties different from the poor items, due to the structure of the features used to determine category membership. This can be demonstrated by the relative ease of imagining what a prototypical good bird looks like, as opposed to a prototypical poor bird.

The distribution of features that gives "good" items salience for humans seems to have no special meaning for the pigeons. During transfer tests, stimulus control always decreased with greater distance on the typicality gradient, quite independent of the point on the human typicality gradient from which the training items were originally selected. The pigeons transferred best to those novel exemplars ranked nearest their training set. This strongly implies that the two groups of pigeons had different representations of the bird category. One possibility we had considered was that the pigeons trained with poor exemplars might extract the features of "birdness" during training and represent the category in a manner similar to the pigeons trained with good exemplars. If this had been the case, they should have responded most vigorously to the good exemplars, a result that would have resembled stimulus intensity dynamism (Hull, 1949). Thus, despite their sensitivity to the structure of the category, that is the intermediate exemplars are located between the good and poor end points, the representation of each category was determined by the exemplars experienced during training. Just as the acquisition experiments described in the first section seem most consistent with an exemplar view of pigeon classification, the training effects of the last two experiments also suggest an exemplar model. The pigeons were more influenced by their previous experience with specific portions of a category than by the particular attributes most often associated with class membership. Thus, the retention of different exemplars would easily explain why each group of pigeons had a different representation of the bird category, and as a byproduct, different patterns of transfer performance.

ANALYSIS OF THE STIMULUS IN BIRD/MAMMAL DISCRIMINATIONS

A serious and recurrent shortcoming in animal categorization experiments has been the scarcity of information about the precise nature of the cues controlling the discrimination. Without this information, inferences about the processes underlying categorization are severely limited. Presented in this third section are the results of a series of experiments examining in detail how pigeons discriminate birds from mammals. Identification of the controlling features should aid in understanding how the pigeons represented the two categories. The strategy we have used resembles a binary search algorithm, with a sequence of tests that attempts to progressively partition the stimulus into smaller parts. The following sections contain brief discussions of the ongoing series of experiments.

Reflection and Rotation. The primary question of this first experiment was whether different orientations of the training stimuli affected performance. If the pigeons used the animal figures alone as the source of the discrimination in the experiments, then a reflection about the vertical axis should have no effect, because the animal would simply be facing the opposite direction. Speculations about reflections around the horizontal axis were not made.

On alternate days of an 8-day test period, half the stimuli of each category were reflected about the vertical axis or were rotated 90° from vertical. Neither the reflection nor the rotation of the stimuli changed performance relative to the items that were not changed during these days.

Figure and Ground. The correspondence between the human typicality ratings and the graded transfer of the pigeons strongly indicates that the animal figures themselves were the source of the discrimination. Nevertheless, because the stimuli were originally intended to be educational, the animal figures are often embedded within the backgrounds of their natural habitats. Some of the birds appear in trees, and some of the mammals on terra firma. Thus, this experiment examined whether or not the pigeons' discriminations depended on background information (Greene, 1983). Four birds and four mammals were chosen from the training set; different birds were selected for the groups trained with "good" and "poor" exemplars, whereas the same mammals were tested for both groups. We devised the test to be as strong as possible by selecting pictures with conspicuous background features or backgrounds that dominated the drawing. These drawings were then altered to make a matched set of items with and without backgrounds and presented as probe trials during a session. The previous contingencies of reinforcement remained in effect. Each item was tested once.

The peck rates for the matched items with and without backgrounds are presented in Fig. 10.7. The peck rates for the two experimental conditions were not significantly different, for either the bird or mammal category ($p > .05$). Removing the background had no reliable effect on the discrimination. The means do show a small change in the direction expected given a degree of stimulus control by the background features; that is, a slightly lower peck rate for the bird trials and slightly higher peck rate on the mammal trials, in the absence of their respective backgrounds. Given that the discrimination was basically not altered by the elimination of the background features, however, suggests that the animal figures themselves were the primary basis for discrimination.

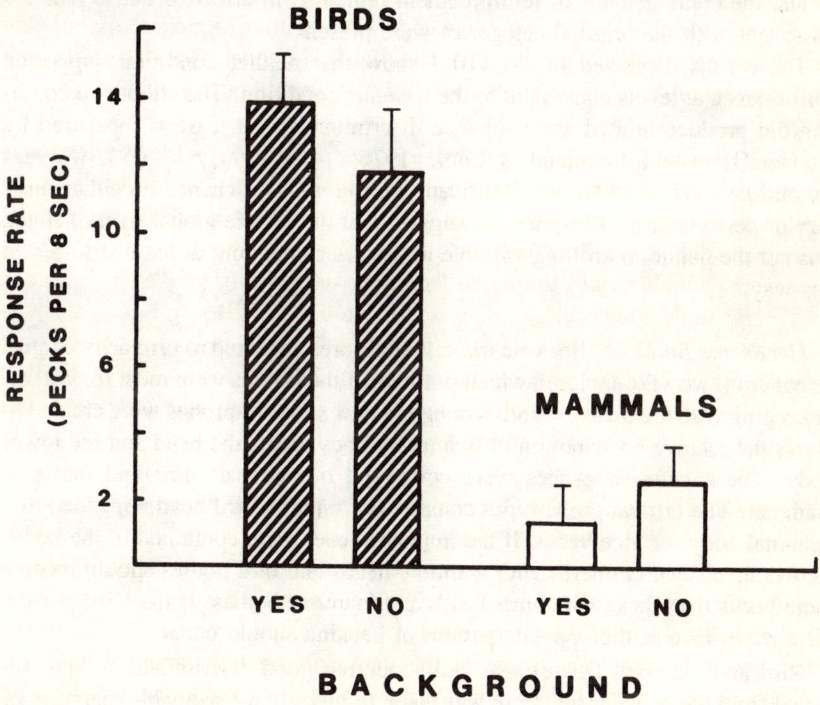

FIG. 10.7. Results of background probe test.

Outlines and Silhouettes. The animal figures can further be divided into two basic components: a set of prominent global features, such as shape or form, and a set of smaller, more detailed local features. If the global features are most important in determining stimulus control, then performance should transfer to test stimuli in which local detailed features have been eliminated. For example, the mammals typically had four prominent legs extending from their bodies, whereas the birds had two less conspicuous legs. If such global features control the discrimination, then simple outlines of the animals should be sufficient to produce the correct response. To test this hypothesis, the local features of the animal figures were removed in two ways. The first eliminated all features and left only the outline of the figures, whereas the second darkened all the features and resulted in a silhouette of the figure.

Four birds and four mammals were selected from the training sets of each group of pigeons, and three matched sets of items were constructed. The *outline* condition was generated from tracing along the boundary of each animal (the backgrounds were also removed from all test figures). The *silhouette* condition was made by blacking in the outlined area. The *control* condition simply had the background removed. Probe items from each category and test condition were tested once in the latter half of four sessions. As in the previous transfer experiments, the contingencies of reinforcement remained in effect, because features consistent with the original categories were present.

The results displayed in Fig. 10.8 show that neither condition supported performance at levels equivalent to the baseline condition. The silhouette condition did produce limited above-chance discrimination, at least as measured by rho (see Herrnstein, Loveland, & Cable, 1976; $t(4) = 2.94$, $p < .05$), whereas the outline condition was not significantly different from chance on either measure of performance. These results suggest that the simple global overall shape was not the major controlling variable in the discrimination, at least as tested in this way.

Heads and Bodies. Because the animal figures appeared to primarily control responding, we next explored which portions of the figures were most important. In keeping with the binary search strategy, a new series of probes were created to assess the relative contribution of two major body areas, the head and the lower body. These *chimera* probes were composed of one part bird and one part mammal. The critical probe types consisted of either a bird head appended to a mammal body, or vice versa. If the important cues were contained in the body, for example, then chimeras with mammal heads and bird bodies should receive more pecks than those with birds heads and mammal bodies. If the head portion were more salient, the opposite pattern of pecking should occur.

Bird and mammal figures that had supported good discrimination were selected from the training set. Care was taken to produce a reasonable marriage of the parts, so that their boundaries matched and the heads and bodies were

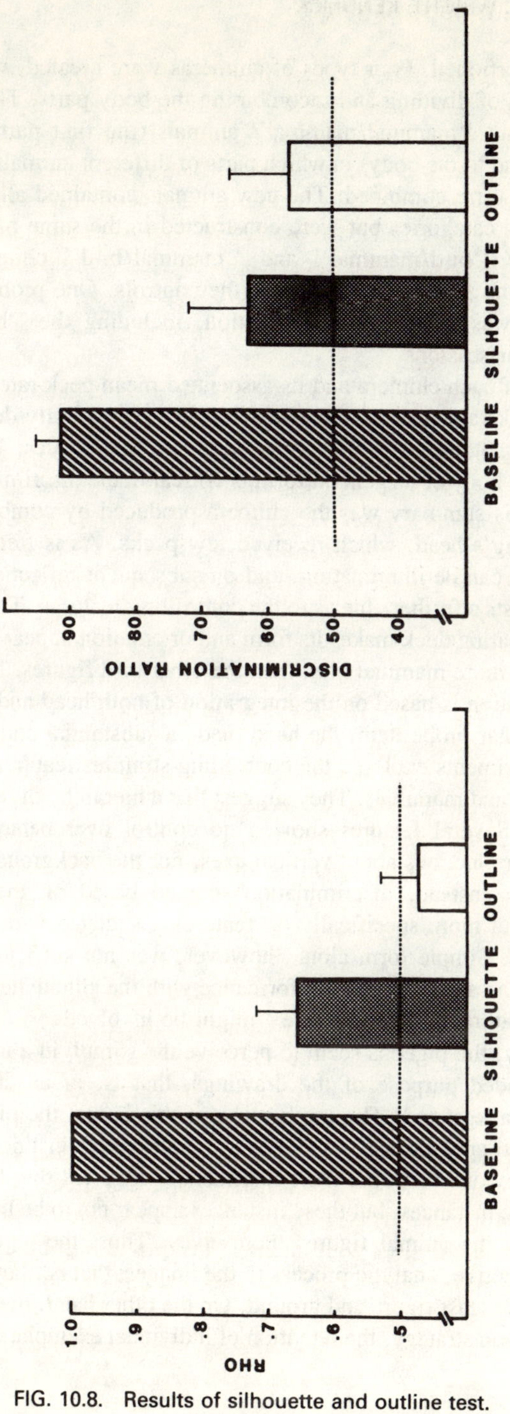

FIG. 10.8. Results of silhouette and outline test.

correctly proportioned. Four types of chimeras were created, with two controlling for the act of dividing and recombining the body parts. These consisted of "bird/bird" and "mammal/mammal" animals (the first part referring to the head, the second to the body) in which parts of different animals from within the same category were combined. The new animals contained all the attributes of their respective categories but were constructed in the same manner as the test chimeras. The "bird/mammal" and "mammal/bird" chimeras were constructed from the same pool of parts as the controls. One probe trial with each chimera type was conducted in extinction, including the "bird/bird" probe trials, over four sessions.

Examples of each chimera and its associated mean peck rate are displayed in Fig. 10.9. Peck rates to 9 of the 10 probes were apparently determined by the body, including 3 of 4 that were composed of conflicting cues. This suggests that a feature of the body or its general form is critical in the discrimination. The only violation of this summary was the chimera produced by combining the duck's body with a pig's head, which received few pecks. As is frequently the case, such "errors" can be illuminating, and on subsequent reflection this particular chimera suggests a further clue as to the controlling features. The overall configuration of a floating duck makes its form and orientation appear more horizontal, and as a result more mammal like, than the other bird figures. It is also possible that discrimination is based on the integration of both head and body parts, and for this particular probe item, the head made a substantial contribution.

These experiments explored the controlling stimulus features in the discrimination of birds and mammals. They suggest that a hierarchy of factors determines performance. Several features showed no control over behavior; neither 90° rotations, nor reflections about vertical axes, nor the background influenced the discrimination. Instead, discrimination seemed based on the animal figures themselves, and more specifically on features associated with the body of the animal figures. Simple form alone, however, was not sufficient to insure discrimination. The above-chance performance with the silhouette test figures suggests that a feature akin to "texture" might be involved.

In summary, the pigeons seem to perceive the stimuli in a manner consistent with the intended purpose of the drawings, that is, of an animal against an independent background. The implication of this is that the pigeons can selectively attend to specific aspects of the stimulus array, in this case the animal figures. In previous sections, the experiments suggested that the pigeons were storing specific instances, but these instances appear not to be the entire stimulus array but only the animal figures themselves. Thus, the pigeons seem to be applying a selective, analytic process to the images; that is, they can decompose the array into at least figure and ground. On the other hand, they appear to use a more nonanalytic strategy, the retention of individual exemplars, to represent the categories.

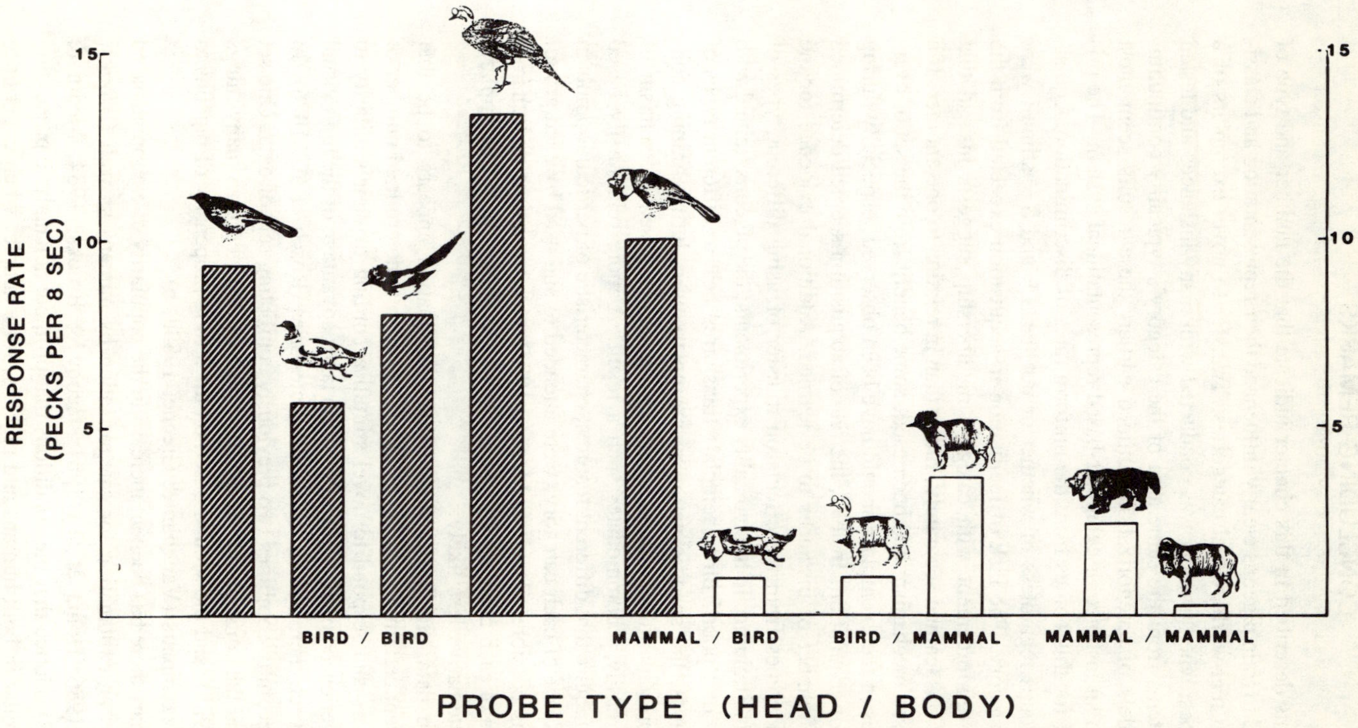

FIG. 10.9. Results of chimera probe test. Examples of each chimera are presented over its data point. The composition of the four probe types is designated so that the first part refers to the category type of the head, whereas the second part refers to the category type of the body. hatched bars represent probe types apparently perceived by the pigeons as birds. The open bars represent probe types apparently perceived as mammals. Only two control mammals were tested because, as in previous experiments, the two groups of pigeons were to receive different examples of birds but the same mammals. Because of the initial results, all chimeras were shown to both groups, requiring the control mammals to be presented twice. This had no noticeable effect.

CONCLUDING REMARKS

The experiments described in this chapter indicate that the number and type of exemplars used to train pigeons clearly influenced their representation and classification of two real-world visual categories. Transfer to novel exemplars of a category was best when training was conducted with many instances and tested with transfer items from the same area of the category's typicality continuum.

Of the models of categorization outlined earlier, these results seem most consistent with an instance or exemplar-based representational system. The primary evidence for this comes from the uniform rates of discrimination acquisition among groups regardless of whether or not the S+ and S− stimuli were unified by a category. The lack of facilitation in acquisition expected from the correlation of reinforcement with category membership suggests instead that each exemplar was learned separately, rather than in relation to one another. If a relationally based mechanism had been used, some benefit from the set of overlapping features in the exemplars should have been observed, especially during the latter part of acquisition. Instead, the rate of acquisition seemed determined simply by the number of instances to be learned. Additional evidence for the exemplar view comes from the effects on transfer of using different types of exemplars during training. In the typicality experiments, the pigeons trained with either "good" or "poor" bird exemplars transferred best to different types of exemplars; that is, the specific nature of exemplars used during training controlled the representation of the category. An exemplar-based mechanism of categorization easily accommodates such a result. A more relationally based system would require that different features, characteristic of particular points in the typicality space of each category, were attended to and used by each group. Whereas this is possible (Rosch & Mervis, 1975), the prominent and characteristic features of each category still are present even in the "poor" exemplars, making this alternative less likely.

Given these data, the most parsimonious explanation appears to be that pigeons represent certain perceptual categories with an undetermined number of stored exemplars and respond to novel stimuli according to their similarity to previously experienced ones. This proposition takes advantage of the high visual similarity between members of basic level categories (Rosch et al., 1976). In addition, it gracefully combines two previously conflicting sets of data. One set centered around the considerable capacity of pigeons to categorize many things (Herrnstein, 1985), and the second based on the strong propensity of pigeons to learn specific exemplars (Vaughan & Greene, 1983a, b).

In recent years there has been an increase in the popularity of exemplar-based mechanisms as an explanation for categorization by humans as well (Brooks, 1978; Estes, 1986; Homa et al., 1981; Jacoby & Brooks, 1984; Medin & Schaffer, 1978). Notice that the conditions that promote nonanalytic processing seem to be similar in both humans and pigeons, for example, when experience with the category is minimal and limited to a few exemplars (Homa et al., 1981), or when the range and distribution of the attributes comprising the category is unknown (Flannagan et al., 1986). It is noteworthy that these conditions describe most of the experiments conducted with animals. An animal confronted with a small set of ill-defined stimuli with which it is unfamiliar might best be served by such an absolute, data-driven, nonanalytic strategy as remembering individual stimuli and their consequences.

The two key components of this exemplar mechanism are a memory for storing instances and a method for comparing the similarity of novel stimuli with the stored instances. Evidence of a good exemplar memory in pigeons is well established. Pigeons have repeatedly demonstrated their ability to remember many specific exemplars and rules (Carter & Werner, 1978; Vaughan & Green, 1983b). Nevertheless, mechanisms for determining similarity remain poorly developed, despite their importance (Lea, 1984). Consider the ''typicality'' gradients produced by pigeons in the present experiments. Some might explain these gradients as simply another example of stimulus generalization. We have no difficulties with this term if it is used as a description of a response gradient in which similarity is prominent. As an explanation it is inadequate, given even the disagreement concerning the mechanisms of stimulus generalization in and of itself (Blough, 1983; Heinemann & Chase, 1975; Riley & Lamb, 1981). Rather than providing an alternative explanation, stimulus generalization may be the same problem as categorization; that is, how do information-processing systems make determinations of similarity (Shepard, 1987)? The clarification of the relation between stimulus generalization and categorization remains a pressing issue.

The evolutionary pressures to minimize processing requirements in a small information-processing system such as the pigeon are self-evident. The abstraction of generalized representations for stimuli and their relations would seem an ideal mechanism for doing so. Yet, pigeons frequently fail to learn such relations in a variety of situations. Aside from not using a more generalized representation in the present categorization experiments, pigeons traditionally have had trouble transferring relations and rules between stimuli in matching or oddity conditional discriminations. Relational learning would clearly reduce the amount of information to be retained, yet despite the presumed advantages of informational integration, there is little persuasive evidence that such analytic processing is a regular occurrence. Some evidence does exist in the conditional discrimination literature suggesting that relational learning can occur in pigeons (Santiago & Wright, 1984; Urcuioli, 1977; Urcuioli & Nevin, 1975; Zentall & Hogan, 1978). Similarly, results showing positive transfer after training with only a single exemplar (Cerella, 1979; Real, Iannazzi, Kamil, & Heinrich, 1984) also suggests that pigeons may establish more generalized representations of categories. It is also fair to say, however, that the vast majority of transfer experiments conducted with pigeons demonstrates a significant bias toward the processing of the absolute values of stimuli rather than their relations. Why pigeons persist in learning if–then rules or retaining specific exemplars remains an intriguing question. One possibility is that the natural world of the pigeon has little need or urgency for extensive relational and analytic cognitive processing. Alternatively, the designing and evolution of relational mechanisms may not be as easy as more absolute, data-driven, exemplar-based systems. Anyone who has programmed a computer appreciates the difficulties involved in generating a truly relational program. A complete understanding of the conditions and use of both analytic and nonanalytic information-processing modes by animals remains to emerge.

ACKNOWLEDGMENTS

This research was supported by a National Research Service Award Postdoctoral Grant from the National Eye Institute (1 FY EY-05653-01A1) to R. G. Cook and a grant from the National Institute of Mental Health (MH-35202) to A. A. Wright. We wish to thank Jackie Rivera for her help in conducting these experiments and Nancy Jacobs for her comments on earlier drafts. R. G. Cook is now at the Department of Psychology, Tufts University, Paige Hall, Medford, MA 02155.

REFERENCES

Ashcraft, M. H. (1978). Property dominance and typicality effects in property statement verification. *Journal of Verbal Learning and Verbal Behavior, 17,* 155–164.

Barsalou, L. W. (1985). Ideals, central tendency, and frequency of instantiation as determinants of graded structure in categories. *Journal of Experimental Psychology: Learning, Memory & Cognition, 11,* 629–649.

Blough, D. S. (1983). Alternative accounts of dimensional stimulus control. In M. L. Commons, R. J. Herrnstein, & A. R. Wagner (Eds.), *Quantitative analyses of behavior: Discrimination processes* (pp. 59–74). Cambridge, MA: Ballinger.

Bourne, L. E., Jr. (1984). Typicality effects in logically defined categories. *Memory & Cognition, 10,* 3–9.

Brooks, L. (1978). Nonanalytic concept formation and memory for instances. In E. Rosch & B. B. Lloyd (Eds.) *Cognition and categorization* (pp. 169–211). Hillsdale, NJ: Lawrence Erlbaum Associates.

Carter, D. E., & Werner, T. J. (1978). Complex learning and information processing by pigeons: A critical analysis, *Journal of the Experimental Analysis of Behavior, 29,* 565–601.

Cerella, J. (1979). Visual classes and natural categories in the pigeon. *Journal of Experimental Psychology: Human Perception & Performance, 5,* 68–77.

Edwards, C. A., & Honig, W. K. (1987). Memorization and "Feature Selection" in the acquisition of natural concepts in pigeons. *Learning and Motivation, 18,* 235–260.

Estes, W. K. (1986). Memory storage and retrieval in category learning. *Journal of Experimental Psychology: General, 115,* 155–174.

Flannagan, M. J., Fried, L. S., & Holyoak, K. J. (1986). Distributional expectations and the induction of category structure. *Journal of Experimental Psychology: Learning, Memory & Cognition, 12,* 241–256.

Greene, S. L. (1983). Feature memorization in pigeon concept formation. In M. L. Commons, R. J. Herrnstein, & A. R. Wagner (Eds.), *Quantitative analyses of behavior: Discrimination processes* (pp. 209–229). Cambridge, MA: Ballinger.

Harter, J. (1979). *Animals.*.New York: Dover Publications.

Heinemann, E. G., & Chase, S. (1975). Stimulus generalization. In W. K. Estes (Ed.), *Handbook of learning and cognitive processes* (Vol. 2). Hillsdale, NJ: Lawrence Erlbaum Associates.

Herrnstein, R. J. (1985). Riddles of natural categorization. In L. Weiskrantz (Ed.), *Animal Intelligence* (pp. 129–142). Oxford: Clarendon Press.

Herrnstein, R. J., & de Villiers, P. A. (1980). Fish as a natural category for people and pigeons. *The Psychology of Learning and Motivation, 14,* 59–95.

Herrnstein, R. J., Loveland, D. H., & Cable, C. (1976). Natural concepts in pigeons. *Journal of Experimental Psychology: Animal Behavior Processes, 2,* 285–311.

Homa, D., Sterling, S., & Treple, L. (1981). Limitations of exemplar-based generalization and the abstraction of categorical information. *Journal of Experimental Psychology: Human Learning and Memory, 7,* 418–439.

Hull, C. L. (1949). Stimulus intensity dynamism (V) and stimulus generalization. *Psychological Review, 56,* 67–76.

Iyari, H. (1979). *Book of animals.* Clinton, NJ: Mayflower Books.

Jacoby, L. L., & Brooks, L. R. (1984). Nonanalytic cognition: Memory, perception, and concept learning. *The Psychology of Learning and Motivation, 18,* 1–47.

Lamb, M. R., & Riley, D. A. (1981). Effects of element arrangement on the processing of compound stimuli in pigeons (*Columba Livia*). *Journal of Experimental Psychology: Animal Behavior Processes, 7,* 45–58.

Lea, S. E. G., & Harrison, S. N. (1978). Discrimination of polymorphous stimulus sets by pigeons. *Quarterly Journal of Experimental Psychology, 30,* 521–537.

Lea, S. E. G. (1984). In what sense do pigeons learn concepts? In H. L. Roitblat, T. G. Bever, & H. S. Terrace (Eds.), *Animal cognition* (pp. 263–277). Hillsdale, NJ: Lawrence Erlbaum Associates.

Malt, B. C., & Smith, E. E. (1982). The role of familiarity in determining typicality. *Memory & Cognition, 10,* 69–75.

Medin, D. L., & Schaffer, M. M. (1978). Context theory of classification learning. *Psychological Review, 85,* 207–238.

Medin, D. L., & Smith, E. E. (1984). Concepts and concept formation. *Annual Review of Psychology, 35,* 113–138.

Mervis, C. B., & Rosch, E. (1981). Categorization of natural objects. *Annual Review of Psychology, 32,* 485–502.

Omohundro, J. (1981). Recognition vs. classification of ill-defined category exemplars. *Memory & Cognition, 9,* 324–331.

Premack, D. (1978). On the abstractness of human concepts: Why it would be difficult to talk to a pigeon. In S. H. Hulse, H. Fowler, & W. K. Honig (Eds.), *Cognitive processes in animal behavior.* Hillsdale, NJ: Lawrence Erlbaum Associates.

Real, P. G., Iannazzi, R., Kamil, A. C., & Heinrich, B. (1984). Discrimination and generalization of leaf damage by blue jays (*Cyanocitta cristata*). *Animal Learning & Behavior, 12,* 202–208.

Riley, D. A., & Lamb, M. R. (1981). Stimulus generalization. In A. D. Pick (Ed.), *Perception and its development.* Hillsdale, NJ: Lawrence Erlbaum Associates.

Rips, L. J., Shoben, E. J., & Smith, E. E. (1973). Semantic distance and the verification of semantic relations. *Journal of Verbal Learning and Verbal Behavior, 12,* 1–20.

Rosch, E. (1975). Cognitive representations of semantic categories. *Journal of Experimental Psychology: General, 104,* 192–233.

Rosch, E. (1978). Principles of categorization. In E. Rosch & B. B. Lloyd (Eds.), *Cognition and categorization* (pp. 27–40). Hillsdale, NJ: Lawrence Erlbaum Associates.

Rosch, E., & Mervis, C. B. (1975). Family resemblances: Studies in the internal structure of categories. *Cognitive Psychology, 7,* 573–605.

Rosch, E., Mervis, C. B., Gray, W. D., Johnson, D. M., & Boyes-Braem, P. (1976). Basic objects in natural categories. *Cognitive Psychology, 8,* 382–429.

Santiago, H. C., & Wright, A. A. (1984). Pigeon memory: *Same/different* concept learning, serial probe recognition acquisition and probe delay effects on the serial-position function. *Journal of Experimental Psychology: Animal Behavior Processes, 10,* 498–512.

Schrier, A. M., Angarella, R., & Povar, M. L. (1984). Studies of concept formation by stumptailed monkeys: Concepts humans, monkeys and letter A. *Journal of Experimental Psychology: Animal Behavior Processes, 10,* 564–584.

Shepard, R. N. (1987). Toward a universal law of generalization for psychological science. *Science, 237,* 1317–1323.

Urcuioli, P. J. (1977). Transfer of oddity-from-sample performance in pigeons. *Journal of the Experimental Analysis of Behavior, 27,* 195–202.

Urcuioli, P. J., & Nevin, J. A. (1975). Transfer of hue matching in pigeons. *Journal of the Experimental Analysis of Behavior, 24,* 149–155.

Uyeda, K. M., & Mandler, G. (1980). Prototypicality norms for 28 semantic categories. *Behavior Research Methods & Instrumentation, 12,* 587–595.

Vaughan, W. Jr., & Greene, S. L. (1983a). Acquisition of absolute discrimination in pigeons. In M. L. Commons, R. J. Herrnstein, & A. R. Wagner (Eds.), *Quantitative analyses of behavior: Discrimination processes* (pp. 231–238). Cambridge, MA: Ballinger.

Vaughan, W. Jr., & Greene, S. L. (1983b). Pigeon visual memory capacity. *Journal of Experimental Psychology: Animal Behavior Processes, 10,* 256–271.

Wasserman, E. A., Kiedinger, R. E., & Bhatt, R. S. (1988). Conceptual behavior in pigeons: Categories, subcategories and psuedocategories. *Journal of Experimental Psychology: Animal Behavior Processes, 14,* 235–246.

Zentall, T. R., & Hogan, D. E. (1978). Same/different concept learning in the pigeon: The effect of negative instances and prior adaptation to transfer stimuli. *Journal of the Experimental Analysis of Behavior, 30,* 177–189.

Author Index

Numbers in *italics* refer to page numbers with full bibliographic information.

Subject Index

D

E

N

O

P